Cerium Oxide Nanocrystals (CeO_{2-x})

Vladyslav Seminko • Pavel Maksimchuk

Cerium Oxide Nanocrystals (CeO_{2-x})

Defects, Luminescence and Redox Activity

Second Edition

Springer

Vladyslav Seminko
Institute for Scintillation Materials
National Academy of Sciences of Ukraine
Kharkiv, Ukraine

Pavel Maksimchuk
Institute for Scintillation Materials
National Academy of Sciences of Ukraine
Kharkiv, Ukraine

ISBN 978-3-032-14129-3 ISBN 978-3-032-14130-9 (eBook)
https://doi.org/10.1007/978-3-032-14130-9

Translation from the Ukrainian language edition: "Дефектна структура, люмінесцентні властивості та антиоксидантна активність нанокристалів оксиду церію (CeO2-x)" by Vladyslav Seminko and Pavel Maksimchuk, © ICMA 2022. Published by Institute for Scintillation Materials of NAS of Ukraine (ICMA). All Rights Reserved.

This Springer imprint is published by the registered company Springer Nature Switzerland AG
The registered company address is: Gewerbestrasse 11, 6330 Cham, Switzerland

Introduction

Cerium oxide (CeO_{2-x}) is known by its ability to form a number of intermediate oxides, the composition of which varies from CeO_2 to Ce_2O_3 as cerium atom can be both tri- and tetravalent in chemical compounds, and cerium ions with both valences (both 4+ and 3+) can be present in the lattice at the same time. The ratio between the contents of ions of different valence is determined by external conditions, in particular, temperature and partial pressure of oxygen in the environment. Thus, by varying the external parameters for cerium oxide crystals or nanocrystals, both the degree of oxygen non-stoichiometry and Ce^{3+}/Ce^{4+} ratio in these materials can be varied. This fact led to the widespread use of these materials as catalysts [1], and in recent years as biological antioxidants capable of protecting living cells from the negative effects of reactive oxygen species [2]. However, the physical mechanism of the antioxidant effect of ceria nanocrystals and its relation to the defect structure of these materials is still under discussion. In general, the main problems that remain relevant today in the study of the role of a defect structure in the formation of the redox activity of cerium oxide nanocrystals were well elucidated in the review article [3] published in a special issue of the journal *Environmental Science: Nano* devoted to the biological action of cerium oxide: "Perhaps most intriguing is the question of the mechanism by which nanoceria appears to beneficially modulate excess reactive oxygen species (ROS). This mechanism or mechanisms needs to be elucidated. Do ROS interact directly with either the plus three or plus four state of the cerium atoms in the lattice (presuming the concept of integer ionic states is even valid) or does all the chemistry occurs in the oxygen vacancies (whose formation may be driven by the reduction of the plus four cerium state). It should also be pointed out that virtually all of our knowledge of the relative concentrations of the cerium states in the lattice comes from very high-energy, in vacuo XPS studies whose relevance to aqueous systems has been questioned. It is well known that these studies tend to overestimate the relative concentration of Ce^{3+} ions. It would be helpful to have an alternative, less potentially damaging analytical technique to obtain an independent measure of this quantity" [3].

In this book, we show that luminescence spectroscopy is such a method, which, on the one hand, opens up the possibility of controlling the processes of interaction of cerium oxide nanocrystals with reactive oxygen species, and on the other hand, makes possible to determine the role of Ce^{3+}/Ce^{4+} ions, oxygen vacancies, and Ce^{3+}-V_0-Ce^{3+} complexes in the formation of antioxidant activity of these nanocrystals. The book discusses in detail the features of the defect structure of cerium oxide nanocrystals, the electronic and optical (in particular, luminescent) properties of both undoped and doped nanocrystals, as well as the role of the defect structure of nanocrystals in the mechanisms of their redox activity in aqueous solutions and biological systems.

References

1. Trovarelli A (1998) Catalysis by ceria and related materials. Imperial College Press, London
2. Huang Y, Ren J, Qu X (2019) Nanozymes: classification, catalytic mechanisms, activity regulation, applications. Chem Rev 119:4357–4412. https://doi.org/10.1021/acs.chemrev.8b00672
3. Reed K, Cormack A, Kulkarni A, Mayton M, Sayle D, Klaessig F, Stadler B (2014) Exploring the properties and applications of nanoceria: is there still plenty of room at the bottom? Environ Sci Nano 1:390–405. https://doi.org/10.1039/C4EN00079J

Contents

Abbreviations

CT	Charge transfer
DMSO	Dimethyl sulfoxide
DPPP	Diphenyl-1-pyrenylphosphine
HMTA	Hexamethylenetetramine
Re	Rare earth ion
ROS	Reactive oxygen species
SOD	Superoxide dismutase
STM	Scanning tunneling microscopy
TEM	Transmission electron microscopy
TSL	Thermostimulated luminescence
UV	Ultraviolet
XPS	X-ray photoelectron spectroscopy
XRD	X-ray diffraction

Chapter 1
Structure and Methods of Obtaining CeO_{2-x} Nanocrystals

Abstract This chapter provides a comprehensive overview of the structural characteristics, defect chemistry, and synthetic approaches to cerium oxide nanocrystals (CeO_{2-x}). Particular emphasis is placed on the role of oxygen non-stoichiometry and the coexistence of Ce^{4+} and Ce^{3+} ions providing high redox activity, oxygen storage capacity, and catalytic performance of nanoceria. The interconnection between vacancy concentration and Ce^{3+}/Ce^{4+} ratio in nanoceria, and size dependence of Ce^{3+} content are discussed. The chapter further outlines the principal synthetic routes including sol-gel, hydrothermal synthesis, colloidal precipitation, and cation exchange methods highlighting their influence on morphology, size, and stoichiometry of nanoparticles. Special attention is given to mixed oxides such as CeO_2-ZrO_2 and CeO_2-Y_2O_3, as well as to doping strategies aimed at tuning the content of oxygen vacancies and Ce^{3+}/Ce^{4+} ratio in ceria nanocrystals. By integrating structural insights with synthesis strategies, this chapter establishes the fundamental principles necessary for rational design of CeO_{2-x} nanomaterials with tailored properties for advanced technological applications.

1.1 Defect Structure of Stoichiometric and Non-stoichiometric Cerium Oxide

Cerium oxide (CeO_2) is currently one of the most studied materials for catalysis [1, 2], photocatalysis [3, 4], energy storage and conversion [5, 6], analysis of gas mixtures [7], etc. One of the key features of cerium oxide is the possibility of coexistence of cerium ions with different valences (Ce^{4+} and Ce^{3+}), due to which the real stoichiometry of cerium oxide corresponds to the formula CeO_{2-x}, where x can take values from 0 to 0.5. The degree of oxygen non-stoichiometry of these materials (and, therefore, the value of x) depends on the temperature and partial pressure of oxygen in the environment, and with increasing temperature and decreasing oxygen pressure, the value of x increases. Due to this, materials based on cerium oxide can effectively store and release oxygen depending on the parameters of environment, which determines the effectiveness of their use for a number of catalytic processes

V. Seminko, P. Maksimchuk, *Cerium Oxide Nanocrystals (CeO_{2-x})*,
https://doi.org/10.1007/978-3-032-14130-9_1

(in particular, three-stage catalysis in automobile converters, obtaining hydrogen by steam conversion methods, etc.), and also as electrolytes for solid-state fuel cells. The transition from bulk samples to nanomaterials makes possible obtaining structures with a high degree of oxygen non-stoichiometry (high content of Ce^{3+} ions) even at room temperatures, which determines the possibility of using cerium oxide nanocrystals in the biological and medical fields. While nanocrystals of cerium oxide (nanoceria) are one of the most promising antioxidant materials, the nature of the antioxidant activity of these materials in biological environments is still the subject of considerable debate [8–10].

Stoichiometric cerium (IV) oxide (CeO_2) has a cubic fluorite-like structure ($Fm2\overline{3}m$ according to the Hermann–Mauguin system or O_h^5 according to the Schönflies system) with the value of the unit cell parameter at room temperature $a = 0.541$ nm (Fig. 1.1). In this structure, each oxygen ion is in the center of a tetrahedron formed by four cerium ions, while each cerium ion is in the center of a cube with the side a/2 ($\approx$ 0.27 nm) formed by eight oxygen ions. In such a structure, cubes of oxygen ions containing cerium ions alternate with empty cubes (such cube is placed in Fig. 1.1 inside a larger cube of cerium ions). The size of the cavity ($\approx$ 0.22 nm) is sufficient to accommodate inside large interstitial ions, which can affect the processes of ion transport in these materials.

The formation of oxygen vacancies in the structure of cerium oxide with temperature increase or decrease in the oxygen partial pressure is accompanied by reduction of part of Ce^{4+} ions to Ce^{3+} ones. Since the ionic radius of the Ce^{3+} ion (0.114 nm) is significantly larger than the ionic radius of the Ce^{4+} ion (0.097 nm), an increase in the degree of oxygen non-stoichiometry leads to an increase in the lattice parameter, which, in turn, leads to decrease in the material density. According to [11], the dependences of the crystal lattice parameter and the density of cerium oxide on the degree of oxygen non-stoichiometry (x) are linear and are described by $a = a_0 + 0.0458 \cdot x$ and $\rho = \rho_0 - 1.286 \cdot x$, respectively, where $a_0 = 0.541$ nm and $\rho_0 = 7.22$ g/cm^3 are crystal lattice parameter and the density of stoichiometric cerium oxide (CeO_2), respectively (Fig. 1.2).

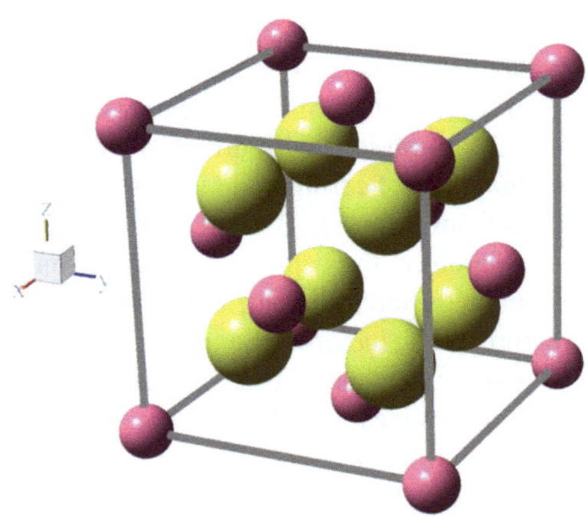

Fig. 1.1 Crystal structure of stoichiometric cerium oxide (CeO_2), cerium ions (Ce^{4+}) are marked as small spheres, oxygen ions (O^{2-}) as large spheres

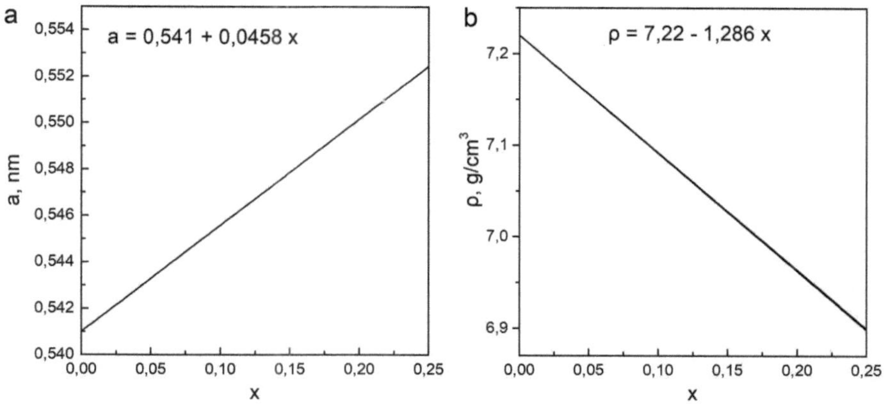

Fig. 1.2 Dependence of the crystal lattice parameter (**a**) and density (**b**) of cerium oxide (CeO_{2-x}) on the degree of oxygen non-stoichiometry for values of x in the range from 0 to 0.25 [11]

The process of oxygen accumulation and release by cerium oxide is determined by the conditions of dynamic equilibrium:

$$CeO_2 \leftrightarrow CeO_{2-x} + x/2O_2 \tag{1.1}$$

A change in the temperature and/or oxygen partial pressure in such a system leads to a change in the equilibrium value of x. Experimental studies and thermodynamic calculations carried out in a number of works [12–14] have shown that an temperature increase and oxygen pressure decrease lead to an increase in the value of the degree of oxygen non-stoichiometry of cerium oxide. In [15], a theoretical model was proposed, which made it possible both to construct the dependence of $x(p, T)$ and to analyze the kinetics of the process of oxygen release from nanocrystals. According to this model, the rate of formation of oxygen vacancies due to oxygen release from the surface of cerium oxide is determined by the rate constants of reduction (k_{red}) and oxidation (k_{ox}), as well as the concentrations of oxygen and oxygen vacancies on the surface of CeO_{2-x} and oxygen in the gas phase. In general, the kinetic equation has the form:

$$\frac{dVo}{dt} = k_{red}\left[O_{CeO2}\right] - k_{ox}\left[Vo\right]\left[O_2\right]^n, \tag{1.2}$$

where $[Vo]$ is the concentration of oxygen vacancies, $[O_{CeO2}]$ is the concentration of oxygen in cerium oxide, $[O_2]$ is the concentration of oxygen in the gas phase. At equilibrium, the rate of formation/neutralization of oxygen vacancies $dVo/dt = 0$.

Since the processes of formation of oxygen vacancies are thermally activated, the rates of reduction (k_{red}) and oxidation (k_{ox}) can be expressed as: $k_{red} = A_{red} \cdot \exp(-E_{red}/kT)$ and $k_{ox} = A_{ox} \cdot \exp(-E_{ox}/kT)$, respectively. Denoting by x_{max} the maximum possible degree of oxygen non-stoichiometry of cerium oxide, and inserting the partial pressure of oxygen p_o into the equation instead of the concentration of oxygen in the gas phase, one can obtain:

$$\frac{x}{x_{max} - x} = \frac{A_{red}}{A_{ox}} p_o^{-n} \exp\left(-\frac{E_{red} - E_{ox}}{RT}\right) \tag{1.3}$$

The analysis of available experimental data [12, 16] allowed the authors [15] to determine the values of the parameters in formula (1.3), namely $x_{max} = 0.35$, $A_{red}/A_{ox} = 8700$, $n = 0.217$, and $E_{red} - E_{ox} = 195.6$ kJ/mole. The dependences of $x(T)$ at different pressure values (from 1 to 10^{-6} bar) according to formula 1.3 are shown in Fig. 1.3. It can be seen from the figure that at temperatures below 1000 °C, the degree of oxygen non-stoichiometry of cerium oxide is negligibly small; on the other hand, the maximum value of x is reached in the temperature range from 2500 to 3500 °C depending on the oxygen pressure in the medium.

The main difference between cerium oxide nanocrystals and corresponding bulk materials is that the degree of oxygen non-stoichiometry of nanocrystals, determined by the value of x, is sufficiently high even at room temperature. This leads to a number of differences in the physical properties of nanocrystals compared to bulk samples.

The transition from bulk crystals to nanocrystals usually leads to a decrease in the unit cell parameter; however, for some oxide nanocrystals (including cerium

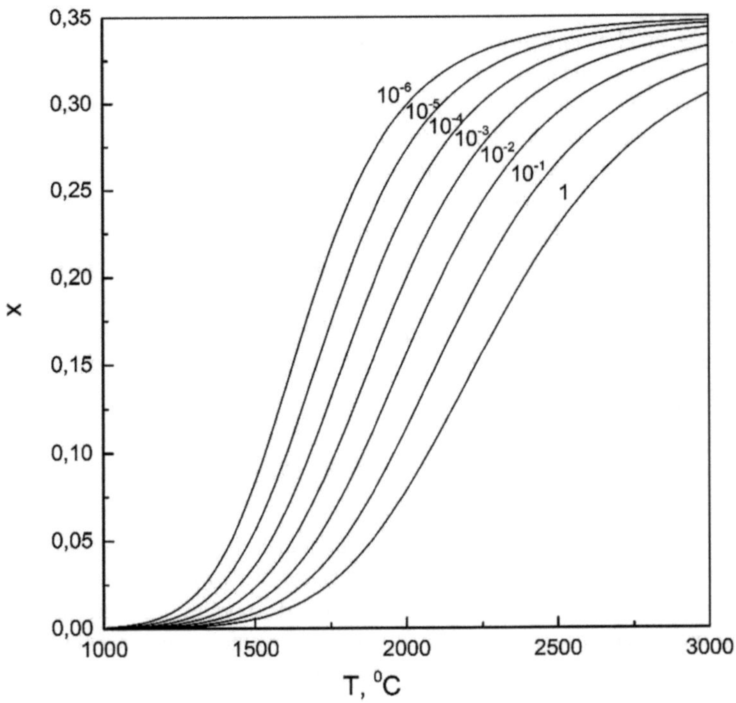

Fig. 1.3 Dependence of the degree of oxygen non-stoichiometry of cerium oxide on the temperature and partial pressure of oxygen in the medium, according to [15], the pressure is indicated in bars (from 1 to 10^{-6})

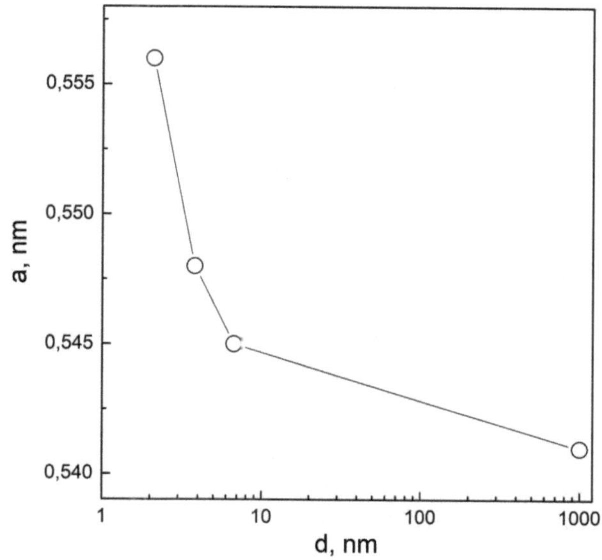

Fig. 1.4 Dependence of the unit cell parameter of CeO_{2-x} nanocrystals on the size of the nanocrystal, experimental data from work [19]

oxide nanocrystals), the opposite pattern was observed, namely an increase in the unit cell parameter with a decrease in the size of the nanocrystal [17–19] (Fig. 1.4).

In [20], it was shown that when the size of CeO_{2-x} nanocrystals was reduced to 1 nm, the unit cell parameter increased by almost 7% compared to bulk crystals preserving the fluorite-like lattice of stoichiometric cerium (IV) oxide (CeO_2).

The increase in the unit cell parameter observed in a number of studies [19, 21] was associated with the presence of a significant number of Ce^{3+} ions in CeO_{2-x} nanocrystals, which ionic radius is larger than for Ce^{4+} ions (the ionic radii of Ce^{4+} and Ce^{3+} are equal to 0.97 Å and 1.14 Å, respectively). This assumption was experimentally confirmed by the results of studies conducted using electron paramagnetic resonance methods (according to which the quota of trivalent Ce^{3+} ions for CeO_{2-x} nanocrystals with a size of 3 nm can reach 18%) [22] and X-ray photoelectron spectroscopy (according to which the quota of Ce^{3+} ions is 44% for 3 nm, 29% for 6 nm, and 17% for 30 nm nanocrystals) [23].

In [24], it was found that when the size of nanocrystals is reduced below a certain value, the reduction of all $Ce^{4+} \rightarrow Ce^{3+}$ ions with the formation of the C-type Ce_2O_3 structure is observed. Such a structure, in contrast to the A-type Ce_2O_3 structure, is not typical for macroscopic cerium oxide, for which the formation of this structure is observed only at temperatures above 2000 °C [25]. The C-type Ce_2O_3 structure is a fluorite-like crystal lattice, similar to the lattice of cerium (IV) oxide (CeO_2) with absent of 25% oxygen ions, and resulting oxygen vacancies arranged along selected directions. The fact that the transition from the stoichiometric cerium (IV) oxide (CeO_2) to the C-type Ce_2O_3 structure does not require the rearrangement of the crystal lattice as a whole is referred by some authors as the main factor that determines the high oxygen capacity and redox activity of small-sized cerium oxide nanocrystals [26]. According to various authors, the formation of this structure takes place at sizes less than 1.5 nm (determined using electron diffraction methods) [19], or less than 3 nm (determined using electron energy loss spectroscopy methods) [24], while according to [20] the formation of this phase was not observed even when the size of the nanocrystal was reduced to 1 nm.

The distribution of Ce^{3+}/Ce^{4+} ions within the volume of CeO_{2-x} nanocrystal was also the subject of a significant number of studies. Thus, in [24], it was shown by the method of electron energy loss spectroscopy that the oxygen non-stoichiometry of nanoparticles increases from the center to the surface. Thus, cerium oxide nanocrystals have a core-shell structure; at the same time, in relatively large particles, the composition of the core is close to stoichiometric CeO_2, and the one of the surface to Ce_2O_3. According to [24], the thickness of the non-stoichiometric surface layer increases sharply with decreasing particle size.

In general, most of the experimental results indicate an abnormally high content of Ce^{3+} ions in CeO_{2-x} nanocrystals associated with a high concentration of oxygen vacancies, which, in turn, cause significant oxygen capacity and catalytic and antioxidant activity of nanocrystalline cerium oxide.

These results, however, found significant criticism in a number of papers [27–29]. Thus, it was found that the effect of increasing the cell parameter with decreasing nanocrystal size is observed not only for cerium oxide and other oxide nanocrystals with ions of variable valence (for example, for titanium oxide TiO_2 nanocrystals [30]), but also for nanocrystals that do not contain ions of variable valence (nanocrystals of zinc oxide ZnO [31], lead titanate $PbTiO_3$ [32], and manganese chromate $MnCr_2O_4$ [33]). As an explanation of the observed effect, the authors of [28] proposed the influence of negative surface tension acting on oxide nanoparticles, and the authors of [34] suggested that the effect is due to the excess Madelung energy arising from the limited size of the nanocrystal. The experimental results show that a high content of Ce^{3+} ions in CeO_{2-x} nanocrystals was criticized in [35]. The methods of X-ray photoelectron spectroscopy, which are the most widely used for determining the content of ions of different valence in CeO_{2-x} nanocrystals, require X-ray irradiation of experimental samples in vacuum chambers, which can lead to the reduction of Ce^{4+} ions to Ce^{3+}. In turn, studies conducted using X-ray absorption spectroscopy methods [36] have shown the absence of Ce^{3+} ions in CeO_{2-x} nanocrystals, which allowed the authors to propose a model of a nanocrystal as an electron sponge, where the electron density is delocalized throughout the volume of the nanocrystal. Thus, to date, there are two opposing views on the issue of the presence or absence of Ce^{3+} ions in cerium oxide nanocrystals as local impurity centers.

Data on the formation energies of defects of various types in stoichiometric and non-stoichiometric cerium oxide differ significantly among different authors [37–41]. In general, most authors note the possibility of formation of the following types of defects in the crystal lattice of cerium oxide:

1. $Ce_{Ce} + 2O_o \leftrightarrow V''''_{Ce} + 2V_o + CeO_2$—simultaneous formation of anionic (oxygen) and cation vacancies (Schottky defect)

2. $Ce_{Ce} \leftrightarrow Ce_i^{\cdot\cdot\cdot\cdot} + V''''_{Ce}$—formation of a cation vacancy (Frenkel defect)
3. $O_o \leftrightarrow O_i'' + V_o$—formation of an anion vacancy according to the Frenkel mechanism

The calculated formation energies of these defects are 3.5 eV, 11.1 eV, and 3.2 eV, respectively [42, 43]. Thus, the most widespread type of defects in cerium oxide should be oxygen vacancies formed by the Frenkel mechanism.

Already at the beginning of the 1960s, it was shown that oxygen vacancies in the structure of cerium oxide formed not random, but ordered structures [44–47]. In later works, a number of possible models for defect ordering in the structure of cerium oxide were proposed.

The most detailed study of the processes of formation and arrangement of oxygen vacancies in the structure of cerium oxide was carried out in recent years using scanning tunneling microscopy (STM) methods [48, 49]. It was shown that treatment at high temperatures leads to the formation of oxygen vacancies on the (111) surface and the tendency of oxygen vacancies to form clusters was revealed. While at short times of high-temperature treatment, only single oxygen vacancies were observed on the (111) surface, and in the subsurface layer, at long times of high-temperature treatment, the most widespread type of defects were linear one-dimensional vacancy clusters.

These results were confirmed in a significant number of subsequent publications [50–54]. Moreover, in [55], it was shown that the presence of one-dimensional vacancy clusters significantly improves the catalytic properties of cerium oxide nanocrystals, in particular, increasing the rate of the $2CO + O_2 \rightarrow 2CO_2$ reaction by almost 2.5 times in the presence of cerium oxide nanoparticles as a catalyst (from 0.21 μmol/g s for nanoparticles with single oxygen vacancies on the surface to 0.51 μmol/g s for nanoparticles in which oxygen vacancies form linear clusters, $t = 160$ °C). Since the catalytic reaction occurs according to the Mars-Van Krevelen mechanism, which includes the formation, migration, and filling of oxygen vacancies on the surface of the nanocrystal, the increase in oxygen mobility due to the presence of one-dimensional vacancy clusters leads to a higher rate of the catalytic reaction [42].

While the authors of [49] note that they did not observe any oxygen diffusion (and, accordingly, no change in the defect structure of the (111) surface) at temperatures lower than 400 °C, other authors [56, 57] show that the transport oxygen takes place even at room temperature. The energy of formation of oxygen vacancies on the surface of nanocrystals is significantly lower than the energy of their formation in the bulk of the crystal (3.2 eV). Thus, for nanocrystals with a size of 2 nm, the calculated formation energy of oxygen vacancies on the (110, 100) surfaces is only 0.4–0.5 eV [58, 59] due to the low coordination of oxygen ions on these surfaces (each of them borders only two, not four, cerium ions). The low energy required for the formation of oxygen vacancies ensures a higher oxygen capacity of cerium oxide nanocrystals compared to their bulk counterparts [60, 61].

The high concentration of oxygen vacancies and the low energy required for their formation even in macroscopic samples of cerium oxide, as well as the ability of this material to accumulate or release significant concentrations of oxygen depending on the partial pressure of oxygen in the environment, led to the widespread use of cerium oxide as an active component of automotive converters capable of removing carbon monoxide, nitrogen oxide, and unburned hydrocarbons from the combustion products of automobile fuel. For automotive converters, the main task is to promote the simultaneous oxidation of carbon monoxide (CO) and hydrocarbons and to reduce the level of nitrogen oxide (NO). Thus, the catalyst must work in a certain range of air-fuel ratios (A/F) to achieve optimal efficiency of all the above-mentioned processes. This, in turn, is achieved by connecting the engine to the electronic control system of the A/F and adding a material for oxygen accumulation, such as cerium oxide [62–64].

An increase in the concentration of oxygen vacancies, and, accordingly, the oxygen capacity of the material can be achieved for mixed oxides, in which part of cerium ions (Ce^{4+}) are replaced by ions that differ in either valence (for example, yttrium ions Y^{3+}, europium Eu^{3+}, terbium Tb^{3+}), or ionic radius (zirconium Zr^{4+} or hafnium Hf^{4+} ions) [65–68]. At the same time, the mechanisms for the formation of additional vacancies in these two cases are of different nature. While the presence of trivalent or divalent ions leads directly to the formation of the corresponding number of oxygen vacancies as charge-compensating defects, the presence of zirconium or hafnium which are isovalent to cerium ions leads to a local distortion of the crystal lattice, as a result of which some of the oxygen ions are weakly associated with the corresponding impurity ions and can easily leave the lattice forming additional oxygen vacancies.

Of all cerium oxide-based mixed oxides, mixed cerium-zirconium (CeO_2-ZrO_2) and cerium-yttrium (CeO_2-Y_2O_3) oxides caused the greatest interest. The use of mixed cerium-zirconium (CeO_2-ZrO_2) oxides together with a metal catalyst (Pt, Rh, Pd) makes it possible to achieve a noticeable decrease in oxygen concentration fluctuations during the transition from reducing to oxidizing conditions and back due to fluctuations in the concentrations of combustion products [69–72].

The local structure of CeO_2-ZrO_2 solid solutions was investigated using the methods of X-ray absorption fine structure spectroscopy (EXAFS spectroscopy) [73]. For CeO_2-ZrO_2 solid solutions of intermediate composition, for example, $Ce_{0.5}Zr_{0.5}O_2$, a significant distortion of the lattice near the zirconium ions was observed. While for cerium oxide, the distance between the cerium ion and the nearest oxygen ion is 0.231 nm, the closest surrounding of the zirconium ion consists of four oxygen atoms at a distance of 0.2115 nm from the Zr^{4+} ion, two atoms at a distance of 0.2324 nm from the Zr^{4+} ion, and the other two of oxygen atoms located at distances greater than 0.26 nm. Longer Zr-O distances for some oxygen ions result in a lower activation energy of oxygen diffusion of these ions. As was shown in [74], the optimal concentration of zirconium ions, at which the oxygen diffusion coefficient is the highest, is about 40 at.%. Due to presence of zirconium ions in mixed cerium-zirconium oxides, oxygen is stored throughout the volume of the material, while in cerium oxide it is limited to the near-surface layer of the material [75]. The oxygen capacity of CeO_2-ZrO_2 also increases in three to four times in the presence of such metals as ruthenium Ru, iridium Ir, palladium Pd, and platinum Pt [76].

Cerium-yttrium oxide (CeO_2-Y_2O_3), in turn, is known as an ionic conductor, which is widely used, in particular, in solid oxide fuel cells (solid oxide fuel cells) as an electrolyte [77, 78]. Data on the optimal concentrations of yttrium in this material differ significantly among different authors, for example, in [79] it was shown that the replacement of part of Ce^{4+} ions by Y^{3+} leads to an increase in the electrical conductivity of the material only at low concentrations of yttrium (up to 1 at.%), while the authors of the paper [80] claim that the optimal concentrations of impurity ions are ~20 at.%.

In general, despite the fact that doping by impurity ions allows to increase the content of oxygen vacancies in the structure of cerium oxide, the effect of the doping on the rate of transport of oxygen ions can be both positive and negative, since impurity ions can bind oxygen vacancies significantly reducing their mobility in the lattice [81–83]. The change of the oxygen transport rate can occur due to capture effects (if impurity ions differ from regular ions in valence) or blocking (if impurity ions differ from regular ions in ionic radius) [6]. The first of these effects is due to the Coulomb attraction of the vacancy and the impurity non-isovalent ion, which leads to an asymmetry of the energy profile and to the complication of migration processes, the second—to a decrease (if the impurity ion is larger than the regular) of free volume, which leads to an increase in the energy barrier for the transition of an oxygen ion to a vacant position near an impurity ion.

The results regarding the effect of impurity ions on the ionic conductivity of cerium oxide differ among different authors. Thus, according to [84], the lowest value of the activation energy of ion transport occurs for doped Sm^{3+} or Gd^{3+} ions (0.75 eV), while for ions with smaller or larger ionic radii, the activation energy increases. In turn, according to the calculations [85], ions with a radius smaller than the radius of the Gd^{3+} ion bind oxygen vacancies which reduce the efficiency of ion transport in the lattice, while ions with a larger ionic radius, on the contrary, push out the oxygen vacancy to an adjacent position. Thus, the influence of impurity ions on the features of the processes of formation and binding of oxygen vacancies in cerium oxide, as well as on the mobility of oxygen in the lattice, remains generally an open question.

1.2 Methods of Obtaining CeO_{2-x} Powders and Nanocrystals

Today, the most widespread methods of obtaining powders and nanocrystals of CeO_{2-x} are methods of colloidal synthesis, sol-gel method, and hydrothermal synthesis.

The Ce^{3+}/Ce^{4+} ratio in CeO_{2-x} nanocrystals obtained in the form of aqueous solutions is determined both by the pH value of these solutions and by the value of the electrochemical potential. To determine the main type of cerium compounds formed at a particular pH value and the redox potential of the environment, Pourbaix diagrams (Fig. 1.5) are usually used, which, in turn, are calculated using the Nernst equation:

$$E = E^0 - \left(\frac{0.0591 \cdot n_{H+}}{n_e}\right)\mathrm{pH} + \left(\frac{0.0591}{n_e}\right)\lg\left(\frac{a_{\mathrm{red}}}{a_{\mathrm{ox}}}\right), \quad (1.4)$$

where E^0 is the standard value of the redox potential (at pH = 0), n_{H+} is the number of H^+ ions involved in the reaction, n_e is the number of electrons involved in the reaction, a_{red} is reductant activity, a_{ox} is oxidant activity.

In turn, the standard value of the redox potential E^0 is determined from the change in Gibbs energy during the corresponding reaction according to the formula:

$$\Delta G^0 = -n_e E^0 F, \quad (1.5)$$

where $F = 96{,}485$ J/V mol is the Faraday constant.

The Pourbaix diagram for the Ce/H_2O system (Fig. 1.5) shows the main types of compounds formed by the interaction of cerium ions with water when pH is varied and/or oxidizing agents/reducing agents are added. The most important from a practical point of view is the pH interval 2.6–8.4, where the equilibrium Ce^{3+}/Ce^{4+} is determined by the chemical formula: $Ce(OH)_4 + 4H^+ + e^- \rightarrow Ce^{3+} + 4H_2O$, and therefore, at $a_{\mathrm{red}} = a_{\mathrm{ox}} = 1$ Nernst equation has the form: $E = 1.98–0.24$ pH. It can be seen from the diagram that the Ce^{3+} ion has greater solubility in water (formation of hydroxides occurs only at pH > 8.4) compared to Ce^{4+} ions (formation of hydroxides and, accordingly, precipitation in the form of a solid phase already at pH = −0.76).

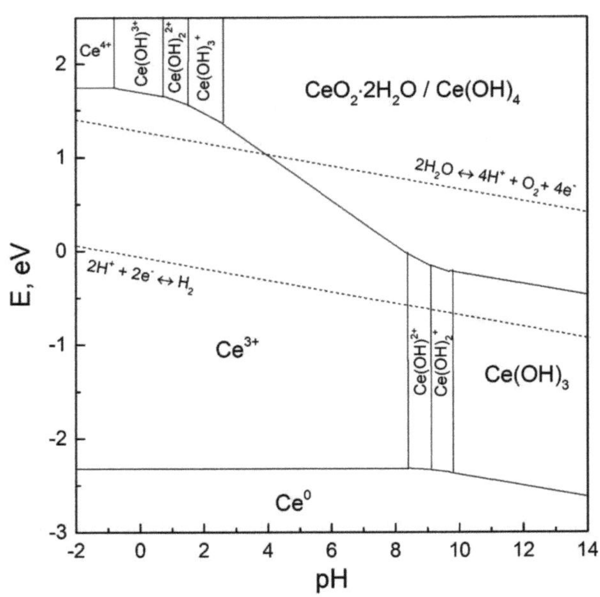

Fig. 1.5 Pourbaix diagram for the Ce/H_2O system [86]

According to theoretical calculations [87, 88], the most stable plane on the surface of CeO_{2-x} nanoparticles is the (111) plane, the (110) plane is less stable, and the (100) plane is even less stable. Thus, the catalytic efficiency for the (100) and (110) planes should be higher compared to the (111) plane. Indeed, in [58], cerium oxide nanocrystals were obtained in the form of rods, the surface of which was formed by (100) and (110) planes, for which the catalytic efficiency of the catalytic oxidation of carbon monoxide (CO) was three times higher than that of nanoparticles for which the surface was bounded by (111) planes. Thus, obtaining cerium oxide nanocrystals with a clearly defined and predetermined morphology allows improving their properties. The process of synthesis of cerium oxide nanocrystals by methods of precipitation or hydrothermal synthesis in aqueous solutions of cerium salts ($Ce(NO_3)_3$, $CeCl_3$ or others) involves the formation of nuclei that grow during the synthesis process. At the same time, both of these stages (nucleation and growth) are controlled by synthesis parameters, such as the temperature and duration of synthesis, as well as the concentrations of the appropriate cerium salt and precipitation agent (NaOH, NH_4OH, etc.). Varying the synthesis parameters allows obtaining nanocrystals of the desired morphology (Table 1.1).

The most promising method of obtaining cerium oxide nanocrystals with a defined morphology is the hydrothermal synthesis method. Thus, in particular, the formation of cerium oxide nanoparticles in the form of polyhedra was shown for the first time in [91]. It was shown that the transition from octahedral nanocrystals (which prevail at lower temperatures) to cubic nanocrystals is achieved due to selective adsorption of nitrate ions NO_3^- at high temperatures on the (100) planes, which leads to a decrease in the free energy for these planes, and, accordingly, to their stabilization. Varying the concentration of the precipitating agent (NaOH) together with the synthesis temperature allowed the authors [92] to obtain nanocrystals in the form of polyhedra ($C = 0.01$–1 M, $t = 100$–180 °C), rods ($C = 1$–9 M, $t = 100$ °C), and cubes ($C = 1$–9 M, $t = 180$ °C). The authors of another publication [93] managed to obtain cerium oxide nanoparticles of four different form factors (octahedral, cubic, rod-like, and grain-like) by varying both the temperature (from 100 to 180 °C) and the type of precipitating agent (Na_3PO_4, NaOH, $CO(NH_2)_2$).

Next, the synthesis methods used to obtain cerium oxide nanocrystals will be presented, the study of which is reported in the following sections.

Table 1.1 The effect of varying the synthesis parameters of cerium oxide nanocrystals in aqueous solutions on the final morphology of nanoparticles [11]

Form of LF	Crystal planes	Growth direction	t, °C	Precipitant concentration	Time, h
Octahedron	{111}	[89]	25–100	<2 M	~12
Truncated octahedron	{111} {100}	[89] [90]	25–100	<2 M	~12
Square rod	{100}	[90]	100–180	>2 M	~12
Cube	{100}	[90]	180–200	>2 M	~24

1.2.1 Synthesis of Cerium Oxide Nanocrystals by the Sol-Gel Method

One of the most accessible methods of synthesis is the so-called sol-gel method [94–97], which is a low-resource chemical synthesis for obtaining finely dispersed crystalline materials. The use of the sol-gel method provides a more complete diffusion of the reacting components, as well as an acceleration of the process of formation of the final product in comparison with solid-phase synthesis.

The use of the sol-gel method in the synthesis of various nanocrystals makes it possible to obtain particles with a small size dispersion, which is possible due to use of precursors that contribute to the formation of gel (for example, solutions of various surface-active substances (surfactants)).

The main advantages of the sol-gel synthesis method are:

- Homogeneity at the molecular level;
- Easy control of the purity of the initial components
- Low agglomeration
- Relatively low process temperature
- Ease of stoichiometry controlling;
- Relatively short times of synthesis

The synthesis of CeO_2 nanocrystals and mixed CeO_2-Re_2O_3 nanocrystals (where Re = Y, Eu) was carried out using one of the varieties of the sol-gel method known as Pechini method [98, 99].

Pechini method, which is mainly used for the synthesis of oxide crystals, is based on the ability of metal chelate complexes (in this case, citrate) to form low molecular weight oligomers with multifunctional alcohols (for example, ethylene glycol). When such a mixture is heated, further polymerization takes place, as well as the formation of a gel, during the decomposition of which metal oxides are formed. Thus, if we assume that the metal complexes are homogeneously distributed throughout the volume in the original mixture and this distribution is preserved after polymerization, there are no reasons why this homogeneity may be disturbed during decomposition. This circumstance makes it possible to use the citrate Pechini method in the work for the synthesis of oxide crystals with uniform distribution of doped ions throughout the particle volume. The method is quite simple in implementation, since, when obtaining the final product, centrifugation, filtration, washing operations are not required, and inorganic metal salts, for example, nitrates or chlorides, are most often used as starting components.

Cerium oxide (IV) CeO_2 (99.99%, Aldrich), yttrium oxide Y_2O_3 (99.99%, Aldrich), europium oxide Eu_2O_3 (99.99%, "Aldrich"), ethylene glycol (1,2-ethanediol –$HOCH_2CH_2OH$) (98%, "Makrokhim"), concentrated nitric acid HNO_3 (63%) ("Macrochem"), citric acid (2-hydroxy-1,2,3-propanetricarboxylic acid ($(HOOCCH_2)_2C(OH)COOH$) (99.0%, "Aldrich"), 35% hydrogen peroxide H_2O_2 ("Macrokhim"), and 25% aqueous solution of ammonia NH_4OH ("Makrokhim") were used as starting components. All starting components were used without additional purification.

Synthesis of samples of CeO_2 nanocrystals consists of several stages. At the initial stage of the synthesis, an aqueous solution of cerium nitrate $Ce(NO_3)_3$ ($c = 0.5$ mol/L) was prepared by dissolving corresponding oxide in a mixture of dilute nitric acid and hydrogen peroxide at a temperature of 60 °C with intensive stirring until the original oxide was completely dissolved, which is also accompanied by the decomposition of excess H_2O_2 and evaporation of excess HNO_3, so the pH value of the $Ce(NO_3)_3$ solution ($c = 0.5$ mol/L) should be about 7.

Then, with intensive stirring, 0.75 g of citric acid dissolved in 1 mL of ethylene glycol was added to 20 mL of the obtained $Ce(NO_3)_3$ solution ($c = 0.5$ mol/L), while metal cations interacted with excess citric acid, i.e., formed citrate complexes according to the chemical reaction:

$$\mathrm{Me}\left(\mathrm{NO}_3\right)_3 + \mathrm{H}_3\,\mathrm{Cit} \rightarrow \mathrm{Me}_n\,\mathrm{H}_{3-n}\,\mathrm{Cit} + \mathrm{nHNO}_3 \quad (1.6)$$

The resulting mixture of precursors was heated for 3 h at a temperature of 80 °C with constant stirring, which led to the evaporation of water and nitric acid formed in the solution according to Eq. (1.6).

Due to removal of water, the solution begins to harden gradually and turns into a gel. The resulting gel is stable for quite a long time (up to several months) and easily forms a homogeneous solution with ethylene glycol in a wide dilution range. This fact allows to adjust the concentration of components and the viscosity of the citrate solution. At the next stage of synthesis, hydrolysis of the gel was carried out. The resulting mixture was neutralized to pH ~ 8 with an aqueous solution of ammonia NH_4OH (10 wt.%). The suspension obtained as a result of the synthesis was dried at a temperature of 110–120 °C. It was then dehydrated (10 h, 250 °C) and annealed (5 h, 550 °C) with intermediate grinding.

Transmission electron microscopy (TEM) of samples obtained by the Pechini method was done using TEM-125 K. Samples for electron microscopy were previously applied to special carbon substrates by dipping and drying of diluted solutions. According to the TEM data, a finely dispersed powder of cerium oxide with an average particle size of about 50 nm was obtained using Pechini method (Fig. 1.6).

The phase composition was monitored using X-ray diffraction analysis (XRD) (Fig. 1.7). According to XRD data, the structure of the obtained samples is characterized by a face-centered cubic lattice with a fluorite structure (JCPDS No. 34-0394) with no additional phases.

The stoichiometry of cerium oxide nanocrystals was controlled by annealing in environments with different reducing properties: oxidizing (air), neutral (argon), and reducing (hydrogen) at a temperature of 1000 °C for 2 h. According to the X-ray diffraction data, the structure of the nanocrystals did not change during the treatment.

The luminescence spectra of cerium oxide nanocrystals at varying high-temperature treatment environments are shown in Fig. 1.8.

After treatment in a neutral or reducing environment, an additional intense luminescence band with a maximum at 390 nm is observed in the spectra of nanocrystals, which, as will be shown in Chap. 2, is related to 5d–4f transitions of Ce^{3+} ions in cerium oxide nanocrystals.

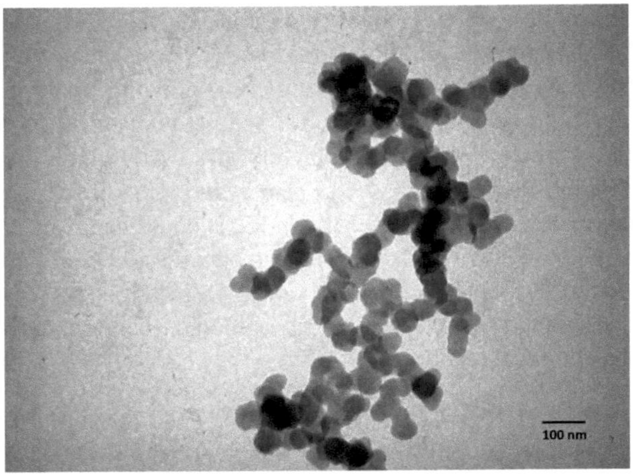

Fig. 1.6 TEM image of CeO_2 nanocrystals synthesized by the Pechini method (average size ~50 nm)

Fig. 1.7 XRD of CeO_2 nanocrystals synthesized by the Pechini method

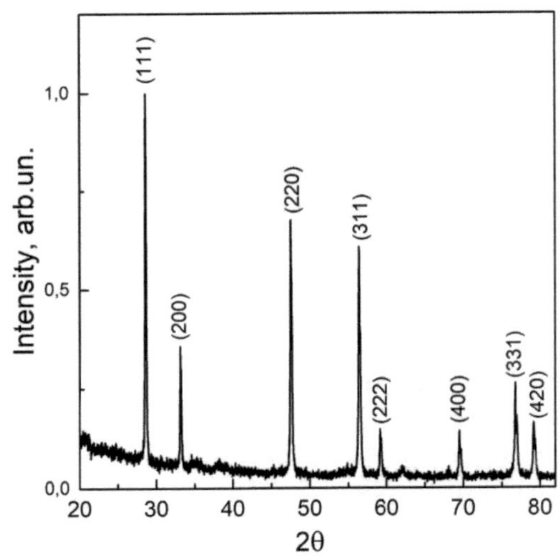

Figure 1.8 shows that while after treatment in an oxidizing medium, the content of Ce^{3+} ions in nanocrystals is negligibly small (because the content of oxygen vacancies is small), the transition to a neutral, and even more, to reducing environment, leads to a significant increase in the content of Ce^{3+} ions, and, accordingly, in the luminescence intensity with a maximum at 390 nm.

Nanocrystals of mixed oxides CeO_2-Re_2O_3 (where Re = Y, Eu) were obtained in the same way as CeO_2 nanocrystals. $Re(NO_3)_3$ solutions (c = 0.5 mol/L) were prepared by dissolving the calculated amount of the corresponding oxide in an aqueous solution of nitric acid with intensive stirring at 40–50 °C. The obtained aqueous solutions of salts were mixed in proportion to the calculated stoichiometric

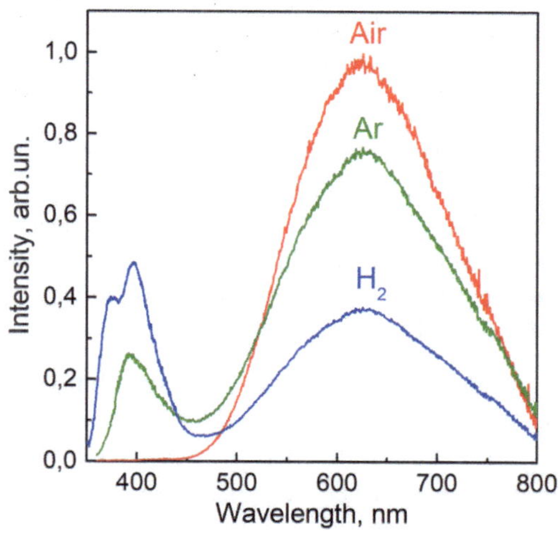

Fig. 1.8 Luminescence spectra of CeO_{2-x} nanocrystals (50 nm), obtained by the sol-gel method, after treatment in oxidizing (1), neutral (2), and reducing (3) environments

coefficients, then a solution of citric acid in ethylene glycol was added and this mixture was kept for 3 h at a temperature of 80 °C with constant stirring. At the next stage of the synthesis, hydrolysis of the citrate gel was carried out with 10 wt.% aqueous ammonia solution, while the pH value of the mixture increased from 7 to 8. The resulting suspension was dried at a temperature of 110–120 °C. Then it was dehydrated (4 h at a temperature of 550 °C) and annealed (10 h at a temperature of 750 °C) with intermediate grinding. As a result, samples of finely dispersed CeO_2-Re_2O_3 powder with a particle size of 50 ± 5 nm were obtained. According to the results of X-ray diffraction, the structure of mixed CeO_2-Re_2O_3 nanocrystals in the entire range of impurity concentrations is characterized by a fluorite-like lattice and the formation of any additional phases is not observed.

Samples of nanocrystals of mixed cerium-zirconium oxides CeO_2-ZrO_2 were synthesized using the sol-gel method using surface-active substances (surfactants).

Ammonium cerium nitrate (IV) $(NH_4)_2Ce(NO_3)_6$ (99.8%, "Aldrich"), zirconium hydrosulfate $Zr(SO_4)_2{\cdot}H_2O$ (99.99%, "Aldrich"), polyethylene glycol-octyl-phenylated OP-10 (98%, "Macrokhim"), and 25% aqueous solution of ammonia NH_4OH ("Macrokhim") were used as starting components. All starting components were used without additional purification.

CeO_2-ZrO_2 nanocrystals were obtained as follows. First, aqueous solutions of $Zr(SO_4)_2$ and $(NH_4)_2Ce(NO_3)_6$ (c = 0.5 mol/L) were prepared by dissolving calculated amounts in distilled water with intensive stirring at room temperature. After that, aqueous solutions of $Zr(SO_4)_2$ and $(NH_4)_2Ce(NO_3)_6$ were mixed in the calculated stoichiometric ratios at room temperature with the addition of an ethanolic solution of OP-10 (surfactant). An aqueous solution of NH_4OH (10 wt. %) was added to the obtained mixture of precursors to pH = 7–8, as a result, a cream-colored precipitate was formed. Then the mixture was heated to 80–90 °C and kept for several hours until a paste-like mass was obtained, which was subjected to high-temperature annealing in air at a temperature of 500 °C for 24 h and at a temperature of 1050 °C for 5 h. As a result, finely dispersed white powders were obtained.

1.2.2 Synthesis of Cerium Oxide Nanocrystals by the Method of Colloidal Coprecipitation

In work, the synthesis of aggregate-stable colloidal solutions of CeO_{2-x} nanocrystals and mixed CeO_2-Re_2O_3 (where Re = Y, Eu, Tb) and CeO_2-ZrO_2 nanocrystals of controlled size was obtained by homogeneous hydrolysis in the presence of hexamethylenetetramine (HMTA) and hydrogen peroxide. It is known that HMTA is subject to slow hydrolysis; therefore, this property is used in the synthesis of oxides of rare earth and transition metals. This technique includes the following stages:

- Hydrolysis of a cerium (III) salt solution in an alkaline environment (pH = 10)
- Adding hydrogen peroxide into the system and boiling the resulting yellow transparent sol with a reflux condenser until decolorization and removal of residual traces of peroxo compounds, nucleation and sedimentation of cerium oxide nanocrystals
- Repeated washing of the sediment from by-products of the reaction (electrolytes) to a neutral value (pH ≈ 7.0) and a negative reaction to the ions that are part of the original salts
- Dispersion of nanoparticles in the presence of an equimolar amount of sodium citrate (citrate ions, being specifically adsorbed on the surface of cerium oxide nanoparticles, form a double electrical layer that prevents agglomeration and coagulation of the sol)
- Stage of dialysis of a colloidal solution

Thus, the proposed method makes possible obtaining nanocrystals of cerium oxide stabilized by sodium citrate with ultra-small sizes (<10 nm), which are stable in water and buffer solutions in a wide range of pH (1–10), and do not show signs of sedimentation during long-term storage.

Cerium (III) chloride pentahydrate $CeCl_3 \cdot 7H_2O$ (99.9%, "Aldrich"), yttrium chloride hexahydrate $YCl_3 \cdot 6H_2O$ (99.99%, "Aldrich"), europium (III) chloride hexahydrate $EuCl_3 \cdot 6H_2O$ (99.99%, Aldrich), terbium (III) chloride hexahydrate $TbCl_3 \cdot 6H_2O$ (99.9%, Aldrich), zirconium hydrosulfate $Zr(SO_4)_2 \cdot H_2O$ (99.99%, Aldrich), hexamethylenetetramine (HMTA, tetraazotricyclo[3,3,1,1]-decane) (98%, "Macrokhim"), 35% hydrogen peroxide H_2O_2 ("Macrokhim"), 25% aqueous solution ammonia NH_4OH ("Macrokhim"), sodium chloride NaCl ("Macrokhim"), and sodium citrate pentahydrate $Na_3Cit \cdot H_2O$ ("Macrokhim") were used as starting components. All starting components were used without additional purification.

According to the method, aqueous solutions of the starting components of the required concentration were first obtained by dissolving the calculated weight in distilled water with intensive stirring at room temperature.

To obtain an aqueous colloidal solution of CeO_2 nanocrystals, 100 mL of $CeCl_3$ solution (with = 0.002 mol/L) is mixed with 100 mL of HMTA solution ($c = 0.04$ mol/L) and stirred for about 3 h at room temperature. After that, 1.8 mL of NH_4OH and 0.6 mL of H_2O_2 are added to the solution, the solution is poured into a reflux flask and heated for 5 h at a temperature of 100 °C. As a result, a colorless transparent solution is obtained.

The resulting solution is evaporated in the flask of a rotary evaporator under vacuum at a bath temperature of 70 °C to 30 mL. NaCl solution was added to the resulting solution ($c = 1$ mol/L) until the solution becomes cloudy (about 2 mL), and the solid phase is precipitated by centrifugation. Then the sediment is separated, and the NaCl solution is added again. The sediment cleaning operation is repeated three times.

After the last stage of centrifugation, a solution of sodium citrate is added to the precipitate in a molar ratio of CeO_{2-x}/Na_3Cit as 1:1. Thus, the solution purified from impurities is transparent, at concentrations of cerium oxide more than 1 g/L it has a weak yellow color and shows opalescence when illuminated from the side.

Then the solution is poured into a dialysis bag, which is placed into a glass filled with distilled water. The dialysis bag is left in the water for 24 h, but after every 6 h, the distilled water in the glass is replaced. As a result of the synthesis, a colorless colloidal solution of cerium oxide nanoparticles is obtained.

The size and structure of obtained CeO_{2-x} nanoparticles were determined using high-resolution transmission electron microscopy (JEOL JEM-2100F (with Schottky-type field emission gun) and JEOL JEM-2100 (with LaB6 filament)), and X-ray diffraction methods (Bruker D8 Discover powder diffractometer, Cu Kα radiation).

The analysis of TEM images (Fig. 1.9a, b) allows determining the average sizes of nanocrystals as 2.6 ± 0.28 nm and 10.3 ± 1 nm. In the following text, we will refer to these nanocrystals as 2 nm and 10 nm ones.

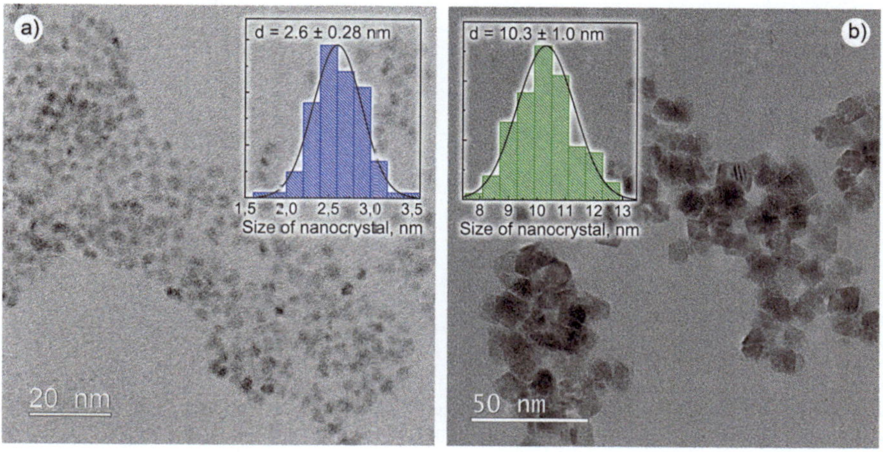

Fig. 1.9 TEM images of CeO_{2-x} nanocrystals obtained by the method of colloidal precipitation from aqueous solutions (**a**—2 nm, **b**—10 nm). Size distributions of nanocrystals are shown in the inserts. (Reprinted with permission from *Journal of Physical Chemistry C*, vol. 127, V. Seminko, P. Maksimchuk, V. Klochkov, Y. Neuhodov, L. Demchenko, S. Yefimova, Reversible CeO_{2-x} and $CeO_{2-x}:Eu^{3+}$ Luminescent Hydrogen Peroxide Sensors with Recovery Rates Controlled by Temperature and UV Irradiation, pp. 10662–10,669, 2023. Copyright 2023 American Chemical Society)

Decrease of the size of nanocrystals leads to an increase of the width of XRD peaks preserving at the same time face-centered cubic structure ($Fm\bar{3}m$) typical for bulk CeO_2. High-resolution TEM images of 2 and 10 nm nanoceria reveal nanocrystals with different orientations. The values of both interplanar (0.324 nm for 2 nm nanocrystals and 0.31 nm for 10 nm) and intraplanar (0.376 nm for 2 nm nanocrystals and 0.36 nm for 10 nm) spacings (Fig. 1.10a, b for 2 nm and 10 nm nanocrystals, respectively) are close to corresponding values for (111) planes in CeO_2 structure. The values of interplanar distances determined from high-resolution TEM are also close to the values of d-spacing for (111) planes determined from XRD (Fig. 1.11) (0.3127 nm). So, as 2 nm, so 10 nm nanoceria expose mainly (111) planes.

Luminescence spectra of colloidal solutions of CeO_{2-x} nanocrystals (2 and 10 nm) contain a broad band with a maximum at 420–430 nm caused by 5d–4f transitions of Ce^{3+} ions in cerium oxide nanocrystals (Fig. 1.12). The nature of this band will be discussed in more detail in Chap. 2. A decrease in the size of nanocrystals from 10 to 2 nm leads to an increase in the intensity of this band, which should be due to an increase in the content of Ce^{3+} ions in nanocrystals.

The synthesis of colloidal solutions of mixed CeO_2-Re_2O_3 (where Re = Y, Eu, Tb) and CeO_2-ZrO_2 nanocrystals was similar to the preparation of colloidal solutions of CeO_2 nanocrystals. However, at the first stage, aqueous solutions of salts ($CeCl_3$, $ReCl_3$, $Zr(SO_4)_4$) were mixed in the necessary stoichiometric ratios.

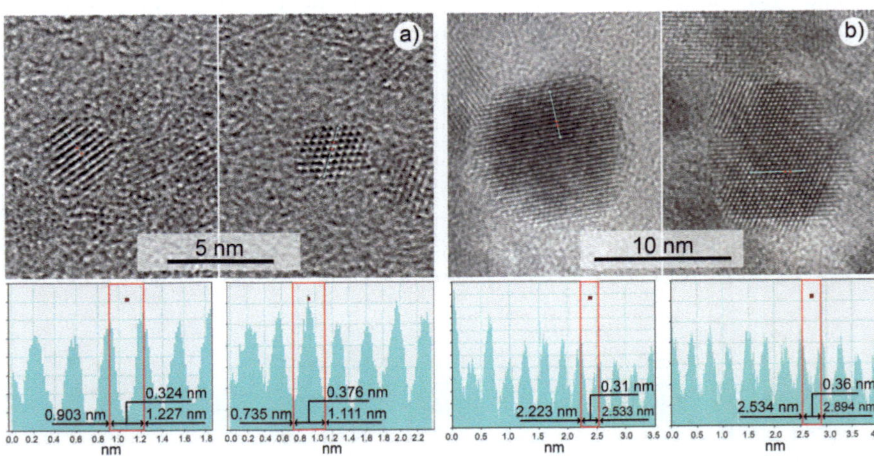

Fig. 1.10 High-resolution TEM images of CeO_{2-x} nanocrystals obtained by the method of colloidal precipitation from aqueous solutions (**a**—2 nm, **b**—10 nm) with intensity profiles for differently oriented nanoparticles. (Reprinted with permission from *Journal of Physical Chemistry C*, vol. 127, V. Seminko, P. Maksimchuk, V. Klochkov, Y. Neuhodov, L. Demchenko, S. Yefimova, Reversible CeO_{2-x} and CeO_{2-x}:Eu^{3+} Luminescent Hydrogen Peroxide Sensors with Recovery Rates Controlled by Temperature and UV Irradiation, pp. 10662–10669, 2023. Copyright 2023 American Chemical Society)

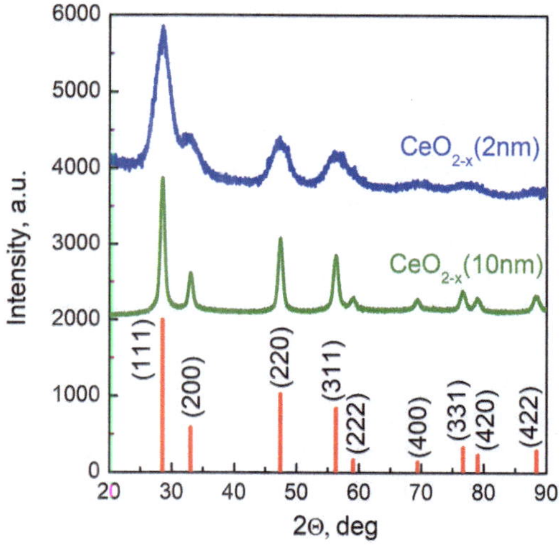

Fig. 1.11 XRD pattern of CeO_2 nanocrystals obtained by the method of colloidal deposition from aqueous solutions (2 nm and 10 nm). (Reprinted with permission from *Journal of Physical Chemistry C*, vol. 127, V. Seminko, P. Maksimchuk, V. Klochkov, Y. Neuhodov, L. Demchenko, S. Yefimova, Reversible CeO_{2-x} and CeO_{2-x}:Eu^{3+} Luminescent Hydrogen Peroxide Sensors with Recovery Rates Controlled by Temperature and UV Irradiation, pp. 10662–10669, 2023. Copyright 2023 American Chemical Society)

Fig. 1.12 Luminescence spectra of CeO_{2-x} nanocrystals (2 and 10 nm) obtained by the method of colloidal coprecipitation

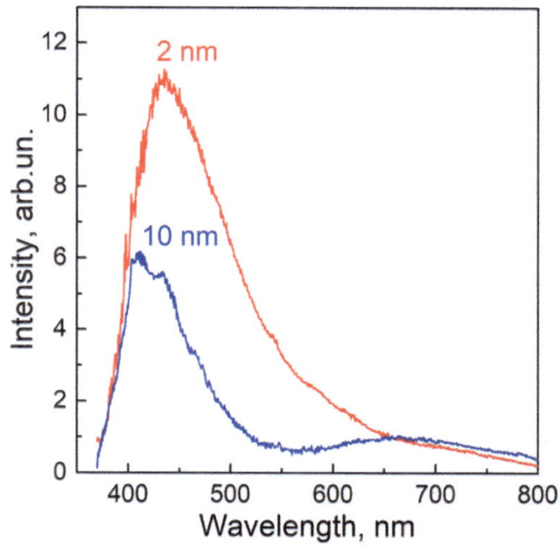

1.2.3 *Doping of Cerium Oxide Nanocrystals by Cation Exchange Method*

The latest achievements in the production of colloidal nanoparticles doped by impurity ions are associated with the development of the cation exchange method [89, 100] based on the ability of various cations to efficiently transfer from aqueous solutions to undoped nanocrystals. This effect was confirmed for a number of nanocrystals of selenides [101], sulfides [102], and tellurides [103]. The cation exchange in oxide nanoparticles has been reported only in a few publications, such as [104], where the authors investigated the process of converting non-magnetic colloidal nanoparticles of copper oxide (CuO) into magnetic nanocrystals of Mn_3O_4, Fe_2O_3, CoO, and NiO by a cation exchange reaction. A relatively small number of studies of cation exchange processes in oxides is associated with smaller Me-O distances in oxide nanocrystals compared to Me-X distances (X = Se, S, Te) in selenide, sulfide, and telluride nanocrystals [89]. In our study, using the methods of luminescence spectroscopy, we found that cation exchange processes can provide effective Re^{3+} (Re = Eu, Tb) doping of cerium oxide nanocrystals in colloidal solutions.

Luminescence spectroscopy methods were used to study the changes in the content of impurity Re^{3+} ions in cerium oxide nanocrystals during the cation exchange process in colloidal solutions. Both investigated Re^{3+} ions (Tb and Eu) reveal characteristic 4f-4f transitions, the spectral position and splitting of which depend on the environment of the impurity ion (in the bulk or on the surface of the nanoparticle). Luminescence spectra for Eu^{3+}- and Tb^{3+}-doped cerium oxide nanocrystals were analyzed earlier in [105, 106]. Tb^{3+} and Eu^{3+} ions have a low concentration quenching of $4f \rightarrow 4f$ luminescence due to an absence of intermediate energy levels between the ^5D_J and ^7F_J multiplets involved in luminescence processes. Therefore, the luminescence intensity of Tb^{3+} and Eu^{3+} ions should be proportional to the concentration of impurity ions.

Luminescence spectra of colloidal solutions of cerium oxide nanocrystals before and 1 h after an addition of terbium chloride ($TbCl_3$) are shown in Fig. 1.13a. The spectra of colloidal solutions of cerium oxide nanocrystals before the addition of

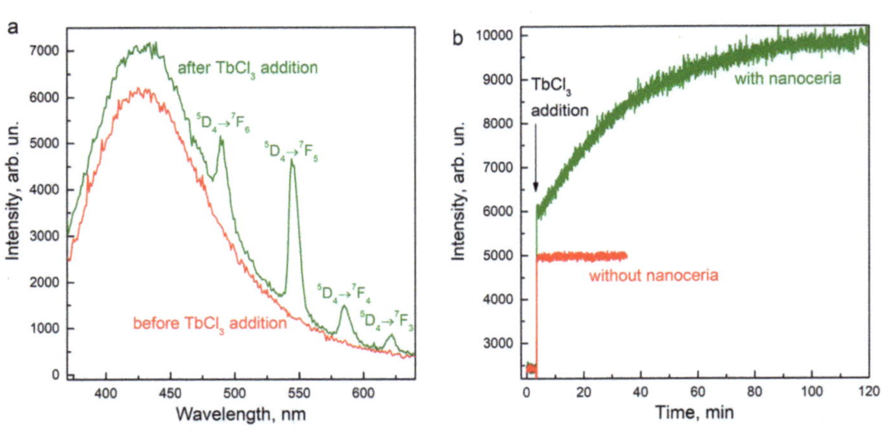

Fig. 1.13 (**a**) Luminescence spectra of colloidal solutions of CeO_{2-x} nanocrystals before and 1 h after an addition of terbium chloride ($TbCl_3$); (**b**) Dynamics of $^5D_0 \rightarrow {}^7F_5$ (545 nm) luminescence of Tb^{3+} ions obtained after adding $TbCl_3$ to aqueous solutions with and without CeO_{2-x} nanocrystals. (Reprinted with permission from *Journal of Luminescence*, vol. 242, V. Seminko, P. Maksimchuk, V. Klochkov, S. Yefimova, Luminescent colloidal ceria nanoparticles doped with RE^{3+} ions (RE = Eu, Tb) via cation exchange mechanism, pp. 118605, 2022. Copyright 2022 Elsevier)

$TbCl_3$ contain a broad band, which, as will be shown in Chap. 2, is associated with 5d → 4f transitions of Ce^{3+} ions. The luminescence spectrum of cerium oxide nanocrystals after the addition of $TbCl_3$, in addition to the Ce^{3+} luminescence band, contains narrow spectral lines at 490 nm, 545 nm, 585 nm, and 622 nm. These spectral lines can be identified as $^5D_4 \rightarrow {}^7F_6$, $^5D_4 \rightarrow {}^7F_5$, $^5D_4 \rightarrow {}^7F_4$ and $^5D_4 \rightarrow {}^7F_3$ transitions of Tb^{3+} ions, respectively. The dynamics of $^5D_4 \rightarrow {}^7F_5$ (545 nm) luminescence of Tb^{3+} ions obtained after adding $TbCl_3$ to aqueous solutions with and without cerium oxide nanocrystals is shown in Fig. 1.13b. For aqueous solutions with cerium oxide nanocrystals, the intensity of Tb^{3+} luminescence gradually increases with time until reaching a maximum approximately 2 h after the addition of $TbCl_3$, while for aqueous solutions without cerium oxide nanocrystals, no further increase in Tb^{3+} luminescence intensity was observed after the moment of adding $TbCl_3$. This slow increase in the intensity of Tb^{3+} luminescence should be attributed to the processes of cation exchange between Tb^{3+} ions from aqueous solutions and cerium ions from CeO_{2-x} nanoparticles. As a result, previously uncoordinated Tb^{3+} ions are fully or partially coordinated in cerium oxide nanocrystals which lead to a gradual increase in the luminescence intensity of Tb^{3+} ions.

The same effect was observed for colloidal solutions of cerium oxide nanocrystals after the addition of europium chloride ($EuCl_3$). The luminescence spectrum of colloidal solutions of cerium oxide nanocrystals 1 h after the addition of $EuCl_3$ (Fig. 1.14a) contains additional spectral lines at 590 nm and 616 nm, which can be attributed to transitions of Eu^{3+} ions: $^5D_0 \rightarrow {}^7F_1$ (magnetodipole) and $^5D_0 \rightarrow {}^7F_2$ (electrodipole), respectively [107–110]. The dynamics of Eu^{3+} luminescence intensity shows the same gradual increase within a few hours after adding $EuCl_3$ to aqueous solutions with cerium oxide, and no increase (except for the initial sharp increase

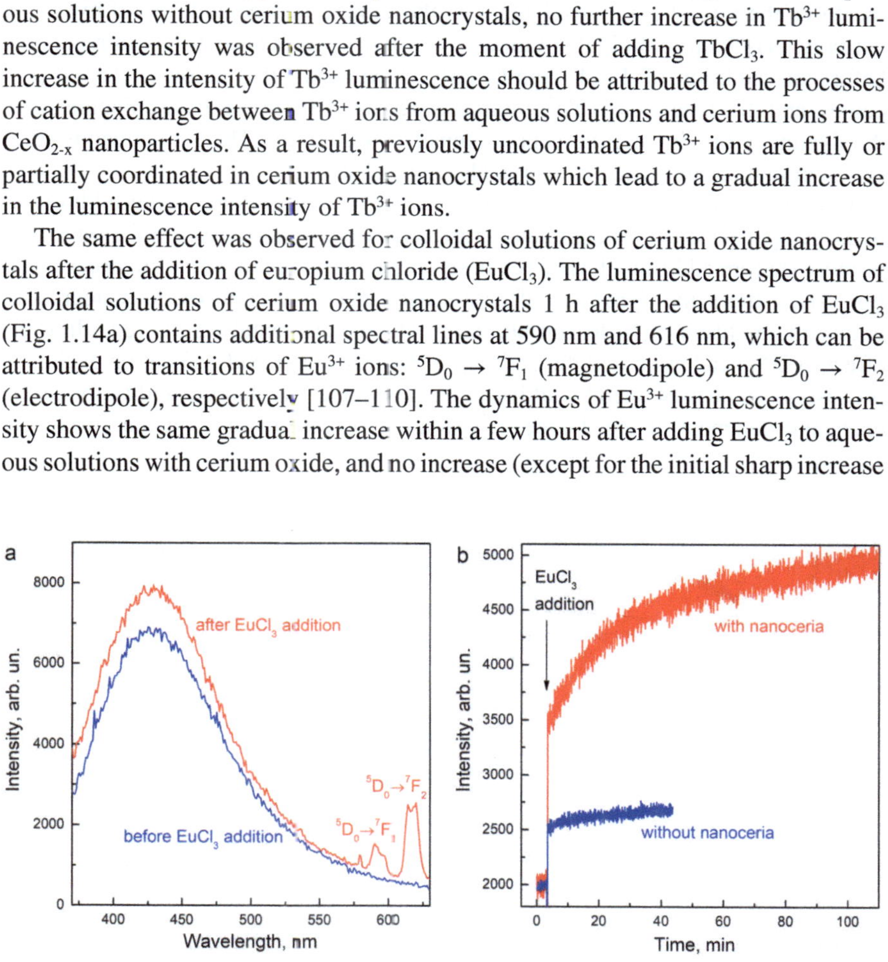

Fig. 1.14 (**a**) Luminescence spectra of colloidal solutions of CeO_{2-x} nanocrystals before and 1 h after adding europium chloride ($EuCl_3$); (**b**) Dynamics of $^5D_0 \rightarrow {}^7F_1$ (590 nm) luminescence of Eu^{3+} ions obtained after adding $EuCl_3$ to aqueous solutions with and without CeO_2 nanocrystals. (Reprinted with permission from *Journal of Luminescence*, vol. 242, V. Seminko, P. Maksimchuk, V. Klochkov, S. Yefimova, Luminescent colloidal ceria nanoparticles doped with RE^{3+} ions (RE = Eu, Tb) via cation exchange mechanism, pp. 118605, 2022. Copyright 2022 Elsevier)

at the time of adding $EuCl_3$) when $EuCl_3$ is added to aqueous solutions without cerium oxide nanocrystals (Fig. 1.14b). A gradual increase in the intensity of Eu^{3+} luminescence after an addition of $EuCl_3$, as well as for Tb^{3+} ions, indicates the processes of cation exchange between Eu^{3+} ions from aqueous solutions and cerium ions from cerium oxide nanoparticles (CeO_{2-x}).

The processes of cation exchange in colloidal solutions of cerium oxide nanocrystals can be affected by the addition of an oxidant or continuous UV irradiation. Figure 1.15a shows the dynamics of Tb^{3+} luminescence intensity after adding $TbCl_3$ to colloidal solutions of cerium oxide nanocrystals in the presence of continuous UV irradiation (325 nm) and without it. Continuous UV irradiation significantly increases the rate of cation exchange. At the same time, the addition of hydrogen peroxide (H_2O_2) to colloidal solutions of cerium oxide nanocrystals immediately before the addition of $TbCl_3$ significantly slows down the cation exchange processes, and the dynamics of Tb^{3+} luminescence intensity in this case coincides with the dynamics of Tb^{3+} luminescence intensity after adding $TbCl_3$ to aqueous solutions without cerium oxide nanocrystals (Fig. 1.15b).

Cation exchange in colloidal solutions of CeO_{2-x} nanocrystals after addition of hydrogen peroxide can be completely restored by continuous UV irradiation for 1 h or more (before addition of $TbCl_3$). These samples, irradiated with ultraviolet radiation, show the same dynamics of cation exchange as cerium oxide nanocrystals before the addition of hydrogen peroxide. The experimental results regarding the effect of UV irradiation and the addition of hydrogen peroxide on cation exchange processes can be explained by taking into account the fact that the ability for cation exchange depends on the degree of oxidation of cerium ions on the surface of

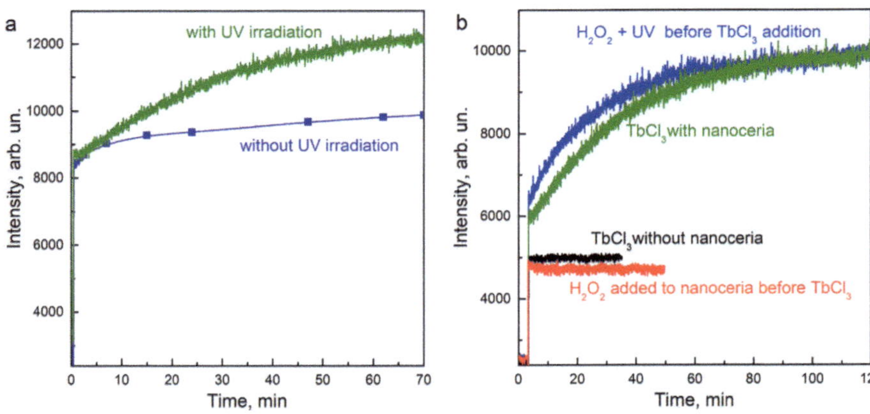

Fig. 1.15 Dynamics of Tb^{3+} luminescence intensity ($\lambda_{reg} = 545$ nm) after adding $TbCl_3$ to colloidal solutions of CeO_{2-x} nanocrystals: (**a**) in the presence of continuous UV irradiation (325 nm) and without it; (**b**) with the addition of hydrogen peroxide to colloidal solutions of CeO_{2-x} nanocrystals before the addition of $TbCl_3$ and with subsequent UV irradiation (1 h, 325 nm). (Reprinted with permission from *Journal of Luminescence*, vol. 242, V. Seminko, P. Maksimchuk, V. Klochkov, S. Yefimova, Luminescent colloidal ceria nanoparticles doped with RE^{3+} ions (RE = Eu, Tb) via cation exchange mechanism, pp. 118605, 2022. Copyright 2022 Elsevier)

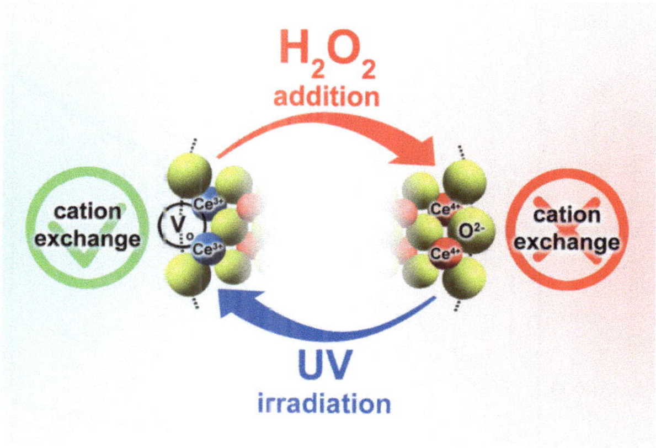

Fig. 1.16 The proposed mechanism of influence of UV irradiation and the addition of oxidants on cation exchange processes in colloidal solutions of CeO_2 nanocrystals. (Reprinted with permission from *Journal of Luminescence*, vol. 242, V. Seminko, P. Maksimchuk, V. Klochkov, S. Yefimova, Luminescent colloidal ceria nanoparticles doped with RE^{3+} ions (RE = Eu, Tb) via cation exchange mechanism, pp. 118605, 2022. Copyright 2022 Elsevier)

nanocrystals. Namely, cation exchange $Re^{3+} \rightarrow Ce^{3+}$ should be much more efficient than $Re^{3+} \rightarrow Ce^{4+}$, which may be related to the difference between the charges and ionic radii of Re^{3+} and Ce^{4+} ions. The addition of hydrogen peroxide leads to the oxidation of $Ce^{3+} \rightarrow Ce^{4+}$, which slows down cation exchange processes, while UV irradiation stimulates the reduction of $Ce^{4+} \rightarrow Ce^{3+}$, which promotes cation exchange.

The general model of the effect of UV radiation and oxidants on cation exchange processes in colloidal solutions of CeO_2 nanocrystals is shown in Fig. 1.16.

The final concentration of Tb^{3+} or Eu^{3+} in cerium oxide nanocrystals is determined both by the content of these ions in the original solution and by the balance between the processes of accumulation and release of ions. Figure 1.17 shows that an increase in the content of Tb^{3+} ions in the colloidal solution of cerium oxide nanocrystals from 50 μM to 5 mM leads to a corresponding increase in the content of impurity ions in CeO_2 nanocrystals. A further increase in the concentration of $TbCl_3$ in the initial solutions does not lead to a higher intensity of luminescence due to a higher content of impurity ions in the cerium oxide nanocrystals, but, instead, is accompanied by the aggregation of nanoparticles; therefore, $C = 5$ mM can be considered as the maximum concentration of $TbCl_3$ at which effective cation exchange can be observed (the maximum initial concentration of $EuCl_3$ is in the same range). Aggregation of colloidal nanoparticles caused by high concentrations of rare earth ions and associated with partial neutralization of the negative surface charge of nanoparticles by positively charged ions of rare earth materials was shown, in particular, in [111], so this effect was expected.

Therefore, the maximum concentrations of doped ions obtained in colloidal nanoparticles by cation exchange are significantly limited by the properties of the colloidal systems themselves. At the same time, simple calculations allow us to estimate that these limiting concentrations are quite high (corresponding to several

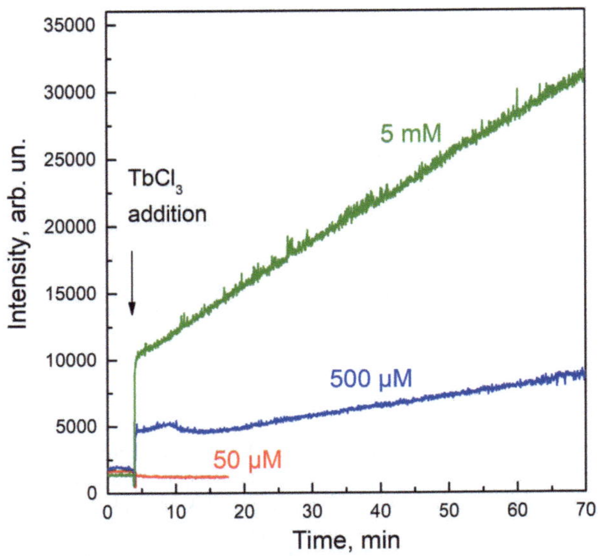

Fig. 1.17 Dynamics of Tb^{3+} luminescence intensity when different concentrations of $TbCl_3$ are added to colloidal solutions of CeO_2 nanocrystals. (Reprinted with permission from *Journal of Luminescence*, vol. 242, V. Seminko, P. Maksimchuk, V. Klochkov, S. Yefimova, Luminescent colloidal ceria nanoparticles doped with RE^{3+} ions (RE = Eu, Tb) via cation exchange mechanism, pp. 118605, 2022. Copyright 2022 Elsevier)

tens of at.% of impurity ions), so the aggregation effect should not create problems for achieving high concentrations of impurity ions in colloidal nanoparticles.

References

1. Montini T, Melchionna M, Monai M, Fornasiero P (2016) Fundamentals and catalytic applications of CeO_2-based materials. Chem Rev 116:5987–6041. https://doi.org/10.1021/acs.chemrev.5b00603
2. Yao HC, Yao YY (1984) Ceria in automotive exhaust catalysts: I. Oxygen storage. J Catal 86:254–265. https://doi.org/10.1016/0021-9517(84)90371-3
3. Zhang YC, Li Z, Zhang L, Pan L, Zhang X, Wang L, Fazal-e-Aleem, Zou J-J (2018) Role of oxygen vacancies in photocatalytic water oxidation on ceria oxide: experiment and DFT studies. Appl Catal B 224:101–108. https://doi.org/10.1016/j.apcatb.2017.10.049
4. Bamwenda GR, Arakawa H (2000) Cerium dioxide as a photocatalyst for water decomposition to O_2 in the presence of Ce^{4+}_{aq} and Fe^{3+}_{aq} species. J Mol Catal A Chem 161:105–113. https://doi.org/10.1016/S1381-1169(00)00270-3
5. Mahato N, Gupta A, Balani K (2012) Doped zirconia and ceria-based electrolytes for solid oxide fuel cells: a review. Nanomater Energy 1:27–45. https://doi.org/10.1680/nme.11.00004
6. Yahiro H, Baba Y, Eguchi K, Arai H (1988) High temperature fuel cell with ceria-yttria solid electrolyte. J Electrochem Soc 135:2077. https://doi.org/10.1149/1.2096212
7. Jasinski P, Suzuki T, Anderson HU (2003) Nanocrystalline undoped ceria oxygen sensor. Sens Actuators B 95:73–77. https://doi.org/10.1016/S0925-4005(03)00407-6

8. Karakoti AS, Monteiro-Riviere NA, Aggarwal R, Davis JP, Narayan RJ, Self WT, Seal S (2008) Nanoceria as antioxidant: synthesis and biomedical applications. JOM 60:33–37. https://doi.org/10.1007/s11837-008-0029-8
9. Lee SS, Song W, Cho M, Puppala HL, Nguyen P, Zhu H, Colvin VL (2013) Antioxidant properties of cerium oxide nanocrystals as a function of nanocrystal diameter and surface coating. ACS Nano 7:9693–9703. https://doi.org/10.1021/nn4026806
10. Grulke E, Reed K, Beck M, Huang X, Cormack A, Seal S (2014) Nanoceria: factors affecting its pro-and anti-oxidant properties. Environ Sci Nano 1:429–444. https://doi.org/10.1039/C4EN00105B
11. Xu Y, Mofarah SS, Mehmood R, Cazorla C, Koshy P, Sorrell CC (2021) Design strategies for ceria nanomaterials: untangling key mechanistic concepts. Mater Horiz 8:102–123. https://doi.org/10.1039/D0MH00554H
12. Zinkevich M, Djurovic D, Aldinger F (2006) Thermodynamic modelling of the cerium–oxygen system. Solid State Ionics 177:989–1001. https://doi.org/10.1016/j.ssi.2006.02.044
13. Hillert M, Jansson B (1986) Thermodynamic model for nonstoichiometric ionic phases—application to CeO_{2-x}. J Am Ceram Soc 69:732–734. https://doi.org/10.1111/j.1151-2916.1986.tb07334.x
14. Ling S (1994) High-concentration point-defect chemistry: statistical-thermodynamic approach applied to nonstoichiometric cerium dioxides. Phys Rev B 49:864. https://doi.org/10.1103/PhysRevB.49.864
15. Bulfin B, Lowe AJ, Keogh KA, Murphy BE, Lubben O, Krasnikov SA, Shvets IV (2013) Analytical model of CeO_2 oxidation and reduction. J Phys Chem C 117:24129–24137. https://doi.org/10.1021/jp406578z
16. Panlener RJ, Blumenthal RN, Garnier JE (1975) A thermodynamic study of non-stoichiometric cerium dioxide. J Phys Chem Solids 36:1213–1222. https://doi.org/10.1016/0022-3697(75)90192-4
17. Spanier JE, Robinson RD, Zhang F, Chan SW, Herman IP (2001) Size-dependent properties of CeO_{2-y} nanoparticles as studied by Raman scattering. Phys Rev B 64:245407. https://doi.org/10.1103/PhysRevB.64.245407
18. Li G, Boerio-Goates J, Woodfield BF, Li L (2004) Evidence of linear lattice expansion and covalency enhancement in rutile TiO_2 nanocrystals. Appl Phys Lett 85:2059–2061. https://doi.org/10.1063/1.1790596
19. Tsunekawa S, Ishikawa K, Li ZQ, Kawazoe Y, Kasuya A (2000) Origin of anomalous lattice expansion in oxide nanoparticles. Phys Rev Lett 85:3440. https://doi.org/10.1103/PhysRevLett.85.3440
20. Hailstone RK, DiFrancesco AG, Leong JG, Allston TD, Reed KJ (2009) A study of lattice expansion in CeO_2 nanoparticles by transmission electron microscopy. J Phys Chem C 113:15155–15159. https://doi.org/10.1021/jp903468m
21. Tsunekawa S, Sivamohan R, Ito S, Kasuya A, Fukuda T (1999) Structural study on monosize CeO_{2-x} nanoparticles. Nanostruct Mater 11:141–147. https://doi.org/10.1016/S0965-9773(99)00027-6
22. Dutta P, Pal S, Seehra MS, Shi Y, Eyring EM, Ernst RD (2006) Concentration of Ce^{3+} and oxygen vacancies in cerium oxide nanoparticles. Chem Mater 18:5144–5146. https://doi.org/10.1021/cm061580n
23. Deshpande S, Patil S, Kuchibhatla SV, Seal S (2005) Size dependency variation in lattice parameter and valency states in nanocrystalline cerium oxide. Appl Phys Lett 87:133113. https://doi.org/10.1063/1.2061873
24. Wu L, Wiesmann HJ, Moodenbaugh AR, Klie RF, Zhu Y, Welch DO, Suenaga M (2004) Oxidation state and lattice expansion of CeO_{2-x} nanoparticles as a function of particle size. Phys Rev B 69:125415. https://doi.org/10.1103/PhysRevB.69.125415
25. Murgida GE, Ferrari V, Ganduglia-Pirovano MV, Llois AM (2014) Ordering of oxygen vacancies and excess charge localization in bulk ceria: a DFT+U study. Phys Rev B 90:115120. https://doi.org/10.1103/PhysRevB.90.115120

26. Skorodumova NV, Simak SI, Lundqvist BI, Abrikosov IA, Johansson B (2002) Quantum origin of the oxygen storage capability of ceria. Phys Rev Lett 89:166601. https://doi.org/10.1103/PhysRevLett.89.166601
27. Sokolov AE, Baranchikov AE, Gil DO, Ivanov VK, Kopitsa GP, Petrunin AA, Fomin EV (2014) On the size effect in nanocrystalline cerium dioxide: is the Tsunekawa model correct? J Surf Investig: X-Ray, Synchrotron Neutron Tech 8:997–1001. https://doi.org/10.1134/S1027451014050413
28. Diehm PM, Agoston P, Albe K (2012) Size-dependent lattice expansion in nanoparticles: reality or anomaly? ChemPhysChem 13:2443–2454. https://doi.org/10.1002/cphc.201200257
29. Qiu L, Liu F, Zhao L, Ma Y, Yao J (2006) Comparative XPS study of surface reduction for nanocrystalline and microcrystalline ceria powder. Appl Surf Sci 252:4931–4935. https://doi.org/10.1016/j.apsusc.2005.07.024
30. Santara B, Giri PK, Imakita K, Fujii M (2014) Microscopic origin of lattice contraction and expansion in undoped rutile TiO_2 nanostructures. J Phys D Appl Phys 47:215302. https://doi.org/10.1088/0022-3727/47/21/215302
31. Ali M, Winterer M (2010) ZnO nanocrystals: surprisingly alive. Chem Mater 22:85–91. https://doi.org/10.1021/cm902240c
32. Akdogan EK, Rawn CJ, Porter WD, Payzant EA, Safari A (2005) Size effects in $PbTiO_3$ nanocrystals: effect of particle size on spontaneous polarization and strains. J Appl Phys 97:084305. https://doi.org/10.1063/1.1872195
33. Bhowmik RN, Ranganathan R, Nagarajan R (2006) Lattice expansion and noncollinear to collinear ferrimagnetic order in a $MnCr_2O_4$ nanoparticle. Phys Rev B 73:144413. https://doi.org/10.1103/PhysRevB.73.144413
34. Perebeinos V, Chan S-W, Zhang F (2002) Madelung model's prediction for dependence of lattice parameter on nanocrystal size. Solid State Commun 123:295–297. https://doi.org/10.1016/S0038-1098(02)00266-1
35. Shcherbakov AB, Zholobak NM, Ivanov VK (2020) Biological, biomedical and pharmaceutical applications of cerium oxide. In: cerium oxide (CeO_2): synthesis, properties and applications. Elsevier, pp 279–358. https://doi.org/10.1016/B978-0-12-815661-2.00008-6
36. Cafun JD, Kvashnina KO, Casals E, Puntes VF, Glatzel P (2013) Absence of Ce^{3+} sites in chemically active colloidal ceria nanoparticles. ACS Nano 7:10726–10732. https://doi.org/10.1021/nn403542p
37. Gauckler LJ, Go M, Schneider D (1997) Nonstoichiometry and defect chemistry of ceria solid solutions. J Electroceram 1:165–172. https://doi.org/10.1023/A:1009928817542
38. Zacherle T, Schriever A, De Souza RA, Martin M (2013) Ab initio analysis of the defect structure of ceria. Phys Rev B 87:134104. https://doi.org/10.1103/PhysRevB.87.134104
39. Filtschew A, Hofmann K, Hess C (2016) Ceria and its defect structure: new insights from a combined spectroscopic approach. J Phys Chem C 120:6694–6703. https://doi.org/10.1021/acs.jpcc.6b00959
40. Grieshammer S, Zacherle T, Martin M (2013) Entropies of defect formation in ceria from first principles. Phys Chem Chem Phys 15:15935–15942. https://doi.org/10.1039/C3CP51913A
41. Wang B, Xi X, Cormack AN (2014) Chemical strain and point defect configurations in reduced ceria. Chem Mater 26:3687–3692. https://doi.org/10.1021/cm500946s
42. Trovarelli A (1998) Catalysis by ceria and related materials. Imperial College Press, London
43. Minervini L, Zacate MO, Grimes RW (1999) Defect cluster formation in M_2O_3-doped CeO_2. Solid State Ionics 116:339–349. https://doi.org/10.1016/S0167-2738(98)00359-2
44. Bevan DJM, Kordis J (1964) Mixed oxides of the type MO_2 (fluorite)-M_2O_3-I oxygen dissociation pressures and phase relationships in the system CeO_2-Ce_2O_3 at high temperatures. J Inorg Nucl Chem 26:1509–1523. https://doi.org/10.1016/0022-1902(64)80038-5
45. Bevan DJM (1955) Ordered intermediate phases in the system CeO_2-Ce_2O_3. J Inorg Nucl Chem 1:49–59. https://doi.org/10.1016/0022-1902(55)80067-X
46. Brauer G, Gingerich KA (1960) Über die oxyde des cers—V: Hochtemperatur-Röntgenuntersuchungen an ceroxyden. J Inorg Nucl Chem 16:87–99. https://doi.org/10.1016/0022-1902(60)80091-7

47. Brauer G, Gradinger H (1954) Über heterotype mischphasen bei seltenerdoxyden. I Z Anorg Allg Chem 276:209–226. https://doi.org/10.1002/zaac.19542760502
48. Jerratsch JF, Shao X, Nilius N, Freund HJ, Popa C, Ganduglia-Pirovano MV, Sauer J (2011) Electron localization in defective ceria films: a study with scanning-tunneling microscopy and density-functional theory. Phys Rev Lett 106:246801. https://doi.org/10.1103/PhysRevLett.106.246801
49. Esch F, Fabris S, Zhou L, Montini T, Africh C, Fornasiero P, Rosei R (2005) Electron localization determines defect formation on ceria substrates. Science 309:752–755. https://doi.org/10.1126/science.1111568
50. Zhang C, Michaelides A, King DA, Jenkins SJ (2009) Oxygen vacancy clusters on ceria: decisive role of cerium f electrons. Phys Rev B 79:075433. https://doi.org/10.1103/PhysRevB.79.075433
51. Wang L, Yu Y, He H, Zhang Y, Qin X, Wang B (2017) Oxygen vacancy clusters essential for the catalytic activity of CeO_2 nanocubes for o-xylene oxidation. Sci Rep 7:1–11. https://doi.org/10.1038/s41598-017-13178-6
52. Ou DR, Mori T, Ye F, Kobayashi T, Zou J, Auchterlonie G, Drennan J (2006) Oxygen vacancy ordering in heavily rare-earth-doped ceria. Appl Phys Lett 89:171911. https://doi.org/10.1063/1.2369881
53. Ye F, Mori T, Ou DR, Cormack AN, Lewis RJ, Drennan J (2008) Simulation of ordering in large defect clusters in gadolinium-doped ceria. Solid State Ionics 179:1962–1967. https://doi.org/10.1016/j.ssi.2008.06.025
54. Torbrügge S, Reichling M, Ishiyama A, Morita S, Custance O (2007) Evidence of subsurface oxygen vacancy ordering on reduced CeO_2 (111). Phys Rev Lett 99:056101. https://doi.org/10.1103/PhysRevLett.99.056101
55. Liu X, Zhou K, Wang L, Wang B, Li Y (2009) Oxygen vacancy clusters promoting reducibility and activity of ceria nanorods. J Am Chem Soc 131:3140–3141. https://doi.org/10.1021/ja808433d
56. Namai Y, Fukui KI, Iwasawa Y (2003) Atom-resolved noncontact atomic force microscopic and scanning tunneling microscopic observations of the structure and dynamic behavior of CeO_2 (111) surfaces. Catal Today 85:79–91. https://doi.org/10.1016/S0920-5861(03)00377-8
57. Namai Y, Fukui K, Iwasawa Y (2003) Atom-resolved noncontact atomic force microscopic observations of CeO_2 (111) surfaces with different oxidation states: surface structure and behavior of surface oxygen atoms. J Phys Chem B 107:11666–11673. https://doi.org/10.1021/jp030142q
58. Zhou K, Wang X, Sun X, Peng Q, Li Y (2005) Enhanced catalytic activity of ceria nanorods from well-defined reactive crystal planes. J Catal 229:206–212. https://doi.org/10.1016/j.jcat.2004.11.004
59. Migani A, Vayssilov GN, Bromley ST, Illas F, Neyman KM (2010) Dramatic reduction of the oxygen vacancy formation energy in ceria particles: a possible key to their remarkable reactivity at the nanoscale. J Mater Chem 20:10535–10546. https://doi.org/10.1039/C0JM01908A
60. Zhang J, Kumagai H, Yamamura K, Ohara S, Takami S, Morikawa A, Suda A (2011) Extra-low-temperature oxygen storage capacity of CeO_2 nanocrystals with cubic facets. Nano Lett 11:361–364. https://doi.org/10.1021/nl102738n
61. Wang D, Kang Y, Doan-Nguyen V, Chen J, Küngas R, Wieder NL, Murray CB (2011) Synthesis and oxygen storage capacity of two-dimensional ceria nanocrystals. Angew Chem Int Ed 50:4378–4381. https://doi.org/10.1002/anie.201101043
62. Fisher GB, Theis JR, Casarella MV, Mahan ST (1993) The role of ceria in automotive exhaust catalysis and OBD-II catalyst monitoring, SAE Technical Paper 931034. SAE International. https://doi.org/10.4271/931034
63. Marecot P, Fakche A, Pirault L, Geron C, Mabilon G, Prigent M, Barbier J (1994) Effect of the preparation procedure on the properties of three-way automotive platinum—rhodium/alumina—ceria catalysts. Appl Catal B 5:43–55. https://doi.org/10.1016/0926-3373(94)00039-5

64. Conesa JC, Martínez-Arias A, Fernández-García M, Soria J (2000) Surface structure and redox chemistry of ceria-containing automotive catalytic systems. Res Chem Intermed 26:103–111. https://doi.org/10.1163/156856700X00138
65. Kim G, Lee N, Kim KB, Kim BK, Chang H, Song SJ, Park JY (2013) Various synthesis methods of aliovalent-doped ceria and their electrical properties for intermediate temperature solid oxide electrolytes. Int J Hydrog Energy 38:1571–1587. https://doi.org/10.1016/j.ijhydene.2012.11.044
66. Wu YC, Lin CC (2014) The microstructures and property analysis of aliovalent cations (Sm^{3+}, Mg^{2+}, Ca^{2+}, Sr^{2+}, Ba^{2+}) co-doped ceria-base electrolytes after an aging treatment. Int J Hydrog Energy 39:7988–8001. https://doi.org/10.1016/j.ijhydene.2014.03.063
67. Vinodkumar T, Rao BG, Reddy BM (2015) Influence of isovalent and aliovalent dopants on the reactivity of cerium oxide for catalytic applications. Catal Today 253:57–64. https://doi.org/10.1016/j.cattod.2015.01.044
68. Eguchi K, Setoguchi T, Inoue T, Arai H (1992) Electrical properties of ceria-based oxides and their application to solid oxide fuel cells. Solid State Ionics 52:165–172. https://doi.org/10.1016/0167-2738(92)90102-U
69. Sugiura M (2003) Oxygen storage materials for automotive catalysts: ceria-zirconia solid solutions. Catal Surv Jpn 7:77–87. https://doi.org/10.1023/A:1023488709527
70. Jen HW, Graham GW, Chun W, McCabe RW, Cuif JP, Deutsch SE, Touret O (1999) Characterization of model automotive exhaust catalysts: Pd on ceria and ceria–zirconia supports. Catal Today 50:309–328. https://doi.org/10.1016/S0920-5861(98)00512-4
71. Bedrane S, Descorme C, Duprez D (2002) Investigation of the oxygen storage process on ceria-and ceria–zirconia-supported catalysts. Catal Today 75:401–405. https://doi.org/10.1016/S0920-5861(02)00089-5
72. Masui T, Ozaki T, Machida KI, Adachi GY (2000) Preparation of ceria–zirconia sub-catalysts for automotive exhaust cleaning. J Alloys Compd 303:49–55. https://doi.org/10.1016/S0925-8388(00)00603-4
73. Vlaic G, Fornasiero P, Geremia S, Kašpar J, Graziani M (1997) Relationship between the zirconia-promoted reduction in the Rh-loaded $Ce_{0.5}Zr_{0.5}O_2$ mixed oxide and the Zr–O local structure. J Catal 168:386–392. https://doi.org/10.1006/jcat.1997.1644
74. Di Monte R, Fornasiero P, Kašpar J, Rumori P, Gubitosa G, Graziani M (2000) Pd/$Ce_{0.6}Zr_{0.4}O_2$/Al_2O_3 as advanced materials for three-way catalysts: part 1. Catalyst characterisation, thermal stability and catalytic activity in the reduction of NO by CO. Appl Catal B 24:157–167
75. Madier Y, Descorme C, Le Govic AM, Duprez D (1999) Oxygen mobility in CeO_2 and $Ce_xZr_{(1-x)}O_2$ compounds: study by CO transient oxidation and $^{18}O/^{16}O$ isotopic exchange. J Phys Chem B 103:10999–11006. https://doi.org/10.1021/jp991270a
76. Martin D, Taha R, Duprez D (1995) Effects of sintering and of additives on the oxygen storage capacity of PtRh catalysts. Stud Surf Sci Catal 96:801–811. https://doi.org/10.1016/S0167-2991(06)81476-X
77. Dudek M, Molenda J (2006) Ceria-yttria-based solid electrolytes for intermediate temperature solid oxide fuel cell. Mater Sci-Pol 24:45–52
78. Fan Z, Chao CC, Hossein-Babaei F, Prinz FB (2011) Improving solid oxide fuel cells with yttria-doped ceria interlayers by atomic layer deposition. J Mater Chem 21:10903–10906. https://doi.org/10.1039/C1JM11550B
79. Wang D, Park D, Griffith J, Nowick A (1981) Oxygen-ion conductivity and defect interactions in yttria-doped ceria. Solid State Ionics 2:95–105. https://doi.org/10.1016/0167-2738(81)90005-9
80. van Herle J, Horita T, Kawada T, Sakai N, Yokokawa H, Dokiya M (1996) Sintering behaviour and ionic conductivity of yttria-doped ceria. J Eur Ceram Soc 16:961–973. https://doi.org/10.1016/0955-2219(96)00012-X
81. Koettgen J, Zacherle T, Grieshammer S, Martin M (2017) Ab initio calculation of the attempt frequency of oxygen diffusion in pure and samarium doped ceria. Phys Chem Chem Phys 19:9957–9973. https://doi.org/10.1039/C6CP04802A

82. Beschnitt S, Zacherle T, De Souza RA (2015) Computational study of cation diffusion in ceria. J Phys Chem C 119:27307–27313. https://doi.org/10.1021/acs.jpcc.5b09499
83. Grieshammer S, Grope BO, Koettgen J, Martin M (2014) A combined DFT+U and Monte Carlo study on rare earth doped ceria. Phys Chem Chem Phys 16:9974–9986. https://doi.org/10.1039/C3CP54811B
84. Andersson DA, Simak SI, Skorodumova NV, Abrikosov IA, Johansson B (2006) Optimization of ionic conductivity in doped ceria. Proc Natl Acad Sci 103:3518–3521. https://doi.org/10.1073/pnas.0509537103
85. Nakayama M, Martin M (2009) First-principles study on defect chemistry and migration of oxide ions in ceria doped with rare-earth cations. Phys Chem Chem Phys 11:3241. https://doi.org/10.1039/B900162J
86. Yu P, Hayes SA, O'Keefe TJ, O'Keefe MJ, Stoffer JO (2006) The phase stability of cerium species in aqueous systems II. The systems. Equilibrium considerations and Pourbaix diagram calculations. J Electrochem Soc 153:C74–C79. https://doi.org/10.1149/1.2130572
87. Sayle TXT, Parker SC, Catlow CRA (1994) The role of oxygen vacancies on ceria surfaces in the oxidation of carbon monoxide. Surf Sci 316:329–336. https://doi.org/10.1016/0039-6028(94)91225-4
88. Conesa J (1995) Computer modeling of surfaces and defects on cerium dioxide. Surf Sci 339:337–352. https://doi.org/10.1016/0039-6028(95)00595-1
89. De Trizio L, Manna L (2016) Forging colloidal nanostructures via cation exchange reactions. Chem Rev 116:10852–10887. https://doi.org/10.1021/acs.chemrev.5b00739
90. Goubin F, Rocquefelte X, Whangbo M-H (2004) Experimental and theoretical characterization of the optical properties of CeO_2, $SrCeO_3$, and Sr_2CeO_4 containing Ce^{4+} (f^0) ions. Chem Mater 16:662–669. https://doi.org/10.1002/chin.200417017
91. Wang ZL, Feng X (2003) Polyhedral shapes of CeO_2 nanoparticles. J Phys Chem B 107:13563–13566. https://doi.org/10.1021/jp036815m
92. Mai HX, Sun LD, Zhang YW, Si R, Feng W, Yan CH (2005) Shape-selective synthesis and oxygen storage behavior of ceria nanopolyhedra, nanorods, and nanocubes. J Phys Chem B 109:24380–24385. https://doi.org/10.1021/jp055584b
93. Wang S, Zhao L, Wang W, Zhao Y, Zhang G, Ma X, Gong J (2013) Morphology control of ceria nanocrystals for catalytic conversion of CO_2 with methanol. Nanoscale 5:5582–5588. https://doi.org/10.1039/C3NR00831B
94. Yamazaki M, Kojima K (1995) Up-conversion fluorescence and electron spin resonance in Er^{3+}-doped CeO_2 heated gel prepared via a sol-gel process. J Mater Sci Lett 14:813–815. https://doi.org/10.1016/j.jcis.2017.02.018
95. Zakharchenya RI (1996) Influence of peptization on the properties of alumina produced from boehmite sols. J Sol-Gel Sci Technol 6:179–186. https://doi.org/10.1007/BF00425975
96. Yoldas BE (1975) Alumina sol preparation from alkoxides. Am Ceram Soc Bull 54:286–288
97. Zhang W, Xie P, Duan C, Yan K, Yin M, Lou L, Krupa JC (1998) Preparation and size effect on concentration quenching of nanocrystalline Y_2SiO_5:Eu. Chem Phys Lett 292:133–136. https://doi.org/10.1016/S0009-2614(98)00656-3
98. Li FB, Li XZ, Hou MF, Cheah KW, Choy WCH (2005) Enhanced photocatalytic activity of Ce^{3+}–TiO_2 for 2-mercaptobenzothiazole degradation in aqueous suspension for odour control. Appl Catal A Gen 285:181–189. https://doi.org/10.1016/j.apcata.2005.02.025
99. Fiorenza R, Bellardita M, D'Urso L, Compagnini G, Palmisano L, Scirè S (2016) Au/TiO_2–CeO_2 catalysts for photocatalytic water splitting and VOCs oxidation reactions. Catalysts 6:121. https://doi.org/10.3390/catal6080121
100. Rivest JB, Jain PK (2013) Cation exchange on the nanoscale: an emerging technique for new material synthesis, device fabrication, and chemical sensing. Chem Soc Rev 42:89–96. https://doi.org/10.1039/C2CS35241A
101. Son DH, Hughes SM, Yin Y, Alivisatos AP (2004) Cation exchange reactions in ionic nanocrystals. Science 306:1009–1012. https://doi.org/10.1126/science.1103755

102. Mukherjee B, Peterson A, Subramanian VR (2012) 1D CdS/PbS heterostructured nanowire synthesis using cation exchange. Chem Commun 48:2415–2417. https://doi.org/10.1039/C2CC16254G
103. Lambert K, Geyter BD, Moreels I, Hens Z (2009) PbTe|CdTe core|shell particles by cation exchange, a HR-TEM study. Chem Mater 21:778–780. https://doi.org/10.1021/cm8029399
104. Ning J, Xiao G, Wang L, Zou B, Liu B, Zou G (2011) Facile synthesis of magnetic metal (Mn, Fe, Co, and Ni) oxides nanocrystals via a cation-exchange reaction. Nanoscale 3:741–745. https://doi.org/10.1039/C0NR00684J
105. Tiseanu C, Parvulescu VI, Sanchez-Dominguez M, Boutonnet M (2012) Temperature induced conversion from surface to "bulk" sites in Eu^{3+}-impregnated CeO_2 nanocrystals. J Appl Phys 112:013521
106. Malleshappa J, Nagabhushana H, Sharma SC, Sunitha DV, Dhananjaya N, Shivakumara C, Nagabhushana BM (2014) Self-propagating combustion synthesis and luminescent properties of nanocrystalline CeO_2:Tb^{3+} (1–10 Mol%) phosphors. J Alloys Compd 590:131–139. https://doi.org/10.1016/j.jallcom.2013.11.213
107. Blasse G (1994) Luminescent materials. Springer-Verlag Telos, Santa Clara
108. Buijs M, Meyerink A, Blasse G (1987) Energy transfer between Eu^{3+} ions in a lattice with two different crystallographic sites: Y_2O_3:Eu^{3+}, Gd_2O_3:Eu^{3+} and Eu_2O_3. J Lumin 37:9–20. https://doi.org/10.1016/0022-2313(87)90177-3
109. Jia G, Wang C, Xu S (2010) Local site symmetry determination of scheelite-type structures by Eu^{3+} spectroscopy. J Phys Chem C 114:17905–17913. https://doi.org/10.1021/jp1077054
110. Cao R, Liu S, Xie L, Pan Y, Cao J, Liu Y (2008) Influence of different site symmetries of Eu^{3+} centers on the luminescence properties of Anderson-based compounds. Inorg Chim Acta 361:2013–2018. https://doi.org/10.1016/j.ica.2007.10.015
111. Grygorova G, Klochkov V, Sedyh O, Malyukin Y (2014) Aggregative stability of colloidal $ReVO_4$: Eu^{3+} (Re= La, Gd, Y) nanoparticles with different particle sizes. Colloids Surf A Physicochem Eng Asp 457:495–501. https://doi.org/10.1016/j.colsurfa.2014.06.024

Chapter 2
Electronic Structure and Luminescent Properties of CeO_{2-x} Nanocrystals

Abstract This chapter discusses the electronic structure and luminescent properties of stoichiometric and non-stoichiometric cerium oxide nanocrystals (CeO_{2-x}). Using stationary and time-resolved luminescence spectroscopy, the effects of oxygen vacancies and Ce^{3+} formation on the optical properties of ceria nanocrystals are shown. $Ce^{4+} \rightarrow Ce^{3+}$ reduction induced by oxygen deficiency leads to the formation of Ce^{3+}-V_0-Ce^{3+} complexes, which are responsible for the characteristic Ce^{3+} 5d–4f luminescence in ceria nanocrystals. The content of Ce^{3+} ions increases as the size of nanoparticles decreases, leading to size dependence in Ce^{3+} luminescence intensity. Doping ceria nanocrystals with Y^{3+} or Zr^{4+} ions enhances oxygen vacancy concentration, further increasing the number of Ce^{3+} optical centers. A generalized energy level scheme is proposed for CeO_{2-x} nanoparticles highlighting the interplay between 4f and 5d states of Ce^{3+} ions and defect-related levels. The results provide critical insights into the mechanisms governing luminescence in CeO_{2-x} nanocrystals and the role of defect engineering in tuning their luminescent properties.

2.1 Luminescence of Stoichiometric and Non-stoichiometric Cerium Oxide Nanocrystals

Cerium oxide is an insulator with a band gap of about 6 eV, the valence band of which is formed by 2p levels of oxygen, and the conduction band by 5d and 6s states of cerium [1]. As was shown before, at 3–3.15 eV above the valence band in the band gap of cerium oxide, there is a narrow sub-band (ranging from 0.5 to 1 eV, according to [1–4]) formed by $4f^0$ states of Ce^{4+} ions. Thus, the optical properties of cerium oxide are determined by charge-transfer transitions from the 2p levels of oxygen to the empty 4f shell of Ce^{4+} ion [2–4].

The presence of an empty $4f^0$ sub-band allows to consider cerium oxide as a semiconductor, for which the $4f^0$ sub-band plays the role of a conduction band providing efficient electron migration during formation of electron-hole pairs at UV irradiation.

V. Seminko, P. Maksimchuk, *Cerium Oxide Nanocrystals (CeO_{2-x})*,
https://doi.org/10.1007/978-3-032-14130-9_2

The transition from stoichiometric to non-stoichiometric cerium oxide nanocrystals significantly changes the electronic properties of nanocrystals, which, in turn, affect the physical characteristics of nanocrystals. The electronic structure of cerium oxide nanocrystals with different degrees of oxygen non-stoichiometry was previously studied in detail from a theoretical point of view using density functional theory [5–8].

These results were obtained using various approximations within the framework of LDA (local density approximation) method with varying Hubbard parameter (Hubbard-U or intrasite electron repulsion). Calculations show that while for stoichiometric cerium oxide in the band gap, there is only an unfilled $4f^0$ sub-band with a width of about 1 eV, and for non-stoichiometric cerium oxide, there is also a band formed by $4f^1$ cerium states, and its position in the band gap is determined by Hubbard parameter [5].

In [9], an interesting hypothesis was put forward regarding the relationship between the features of the processes of formation of oxygen vacancies in cerium oxide and the quantum effect of electron localization. This effect was previously shown for metallic cerium, for which $\gamma \rightarrow \alpha$ isostructural transition is observed, when under a certain pressure, the volume of the unit cell of the material sharply decreases while maintaining the same type of crystal structure [10]. To explain this effect, the authors [11, 12] have used the model of the Mott transition (metal-insulator transition), which is accompanied by the delocalization (metallization) of the 4f electron.

A similar effect, according to [9], is also observed for cerium oxide during formation of oxygen vacancies. The authors [9] have shown that the process of formation of oxygen vacancies in cerium oxide is significantly facilitated due to localization of two electrons remained during release of oxygen on the neighboring Ce^{4+} ions with their transition to Ce^{3+} ones.

As a result, Ce^{3+}-Vo-Ce^{3+} complexes are formed (where Vo is an oxygen vacancy). The calculations [9] have shown that while the formation of a separate oxygen vacancy requires about 4.55 eV, the formation of the Ce^{3+}-Vo-Ce^{3+} complex needs only about 0.26 eV. This effect can explain the high content of oxygen vacancies in cerium oxide and its high oxygen capacity, as well as the formation of the C-type Ce_2O_3 structure in cerium oxide nanocrystals (i.e., the structure similar to the structure of CeO_2 with 25% oxygen vacancies). A similar modification of the electronic properties of cerium oxide with an increase in the degree of non-stoichiometry should be also manifested in the optical properties of this material. In particular, it can be expected that the luminescence spectra of non-stoichiometric cerium oxide should contain a band corresponding to the d $\rightarrow$ f transitions of cerium ion, the intensity of which should depend significantly on the degree of non-stoichiometry. However, such luminescence transitions in cerium oxide were not observed before, and the publications (the number of which remains relatively small) devoted to the luminescence of cerium oxide nanocrystals were mainly focused on the luminescence of impurity ions and defects (including F centers) [13].

The location of oxygen vacancies relative to trivalent impurity ions (Eu^{3+}, Sm^{3+}, Pr^{3+}) was also investigated in a number of papers [14, 15, 16]. It was shown in [17] that for rare earth ions with a radius smaller than the radius of the Gd^{3+} ion, the vacancy is localized in the first coordination sphere of the impurity ion, while for ions with a larger radius, the vacancy is repelled to the second coordination sphere. At the same time, the radius of the Ce^{3+} ion is bigger than the radius of the Gd^{3+} ion, and thus, according to the results of [17], one could expect that the vacancy will be repelled into the second coordination sphere. However, as it was mentioned above [9], the formation of oxygen vacancies in the CeO_2 crystal matrix requires the least amount of energy if it is accompanied by the localization of electrons on two Ce^{3+} ions in the immediate vicinity of the vacancy.

To study the features of the defect and electronic structure of cerium oxide nanocrystals with different degrees of oxygen non-stoichiometry, we used the methods of stationary and time-resolved luminescence spectroscopy. As was noted in Chap. 1, varying the degree of oxygen non-stoichiometry of nanocrystals affects their defect structure, which, in turn, should change the electronic properties of nanocrystals. In this work, in order to obtain cerium oxide nanocrystals with different degrees of oxygen non-stoichiometry, high-temperature annealing of sol-gel synthesized nanocrystals (50 nm) was carried out in different environments (oxidizing (air), neutral (argon), and reducing (hydrogen)).

Luminescence spectra of nanocrystals after treatment in various environments are shown in Fig. 2.1.

Luminescence spectra of cerium oxide nanocrystals after treatment in oxidizing environment (Fig. 2.1a) at $T = 77$ K contain only a band with a maximum at 620 nm. Increasing the temperature to 293 K leads to the temperature quenching of this band. A significant band width (4000 cm^{-1}), a large Stokes shift (13,100 cm^{-1}), as well as a strong temperature dependence make it possible to assume that this luminescence band can be caused by radiative relaxation in Ce^{4+}–O^{2-} charge-transfer complexes.

After treatment in neutral or reducing environment (Fig. 2.1b, c), an additional intense luminescence band with a maximum at 390 nm and width of 2500 cm^{-1} is observed in the spectrum. As the temperature decreases, the intensity of this band increases not as noticeably as of the 620 nm band (the integral intensity of the 620 nm band increases in 5.5 times as the temperature decreases, and the one of the 390 nm band only in 1.7 times).

The luminescence band with a maximum at 390 nm at $T = 77$ K shows splitting into two components, the distance between which is about 1900 cm^{-1} (Fig. 2.1c). This splitting is typical for the ground state of Ce^{3+} ion (^2F), which is split into two components ($^2F_{5/2}$ and $^2F_{7/2}$) as a result of the spin-orbit interaction, so the band at 390 nm can be attributed to 5d–4f luminescence of Ce^{3+} ions in CeO_2 nanocrystals. This assumption is confirmed by the analysis of the luminescence decay curves (Fig. 2.2) taken in this band. The luminescence decay times for the band at 390 nm are in the nanosecond range (15 ns at $T = 293$ K and 25 ns at $T = 77$ K), which is also typical for d-f transitions of rare earth ions [18]. Unlike the band at 390 nm, the decay times of the band with a maximum at 620 nm are in the microsecond range.

The fact that the band with a maximum at 390 nm is due to the 5d–4f luminescence of Ce^{3+} ions is also confirmed by the significant dependence of the intensity

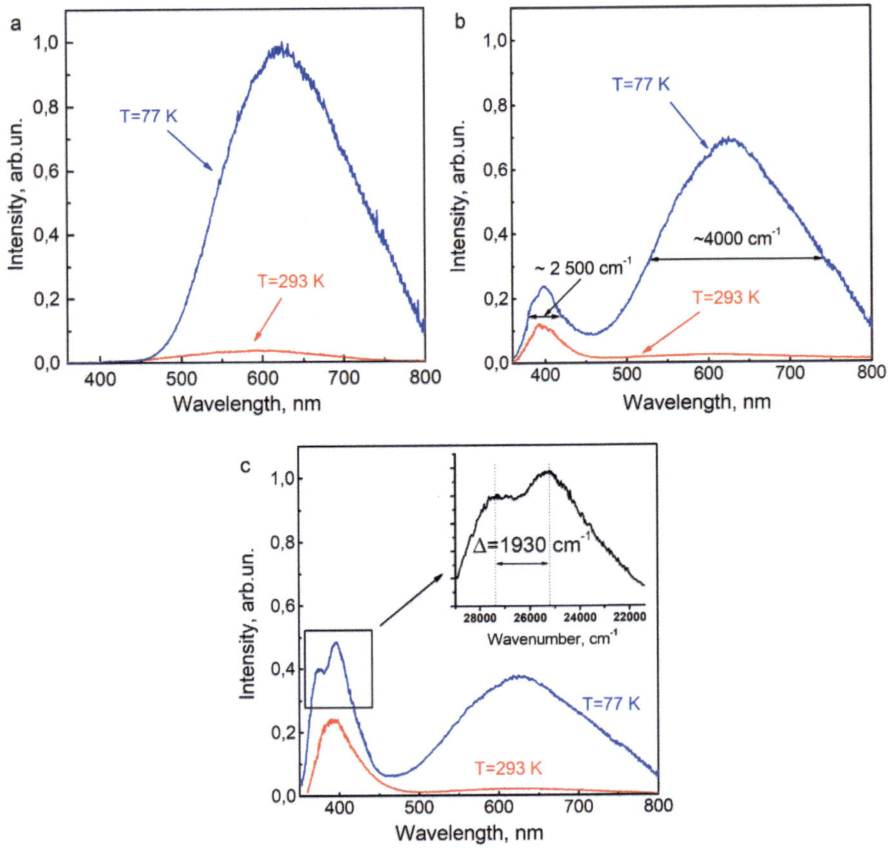

Fig. 2.1 Luminescence spectra of CeO_{2-x} nanocrystals (50 nm) after high-temperature (1000 °C) treatment in different environments (**a**—oxidizing, **b**—neutral, **c**—reducing), λ_{exc} = 325 nm. The inset shows the splitting of the luminescence band with a maximum at 390 nm (in cm^{-1}). (Reprinted with permission from *Journal of Luminescence*, vol. 145, A. Masalov, O. Viagin, P. Maksimchuk, V. Seminko, I. Bespalova, A. Aslanov, Yu. Malyukin, and Yu. Zorenko, Formation of luminescent centers in CeO_2 nanocrystals, pp. 61–64, 2014. Copyright 2014 Elsevier)

of this band on the environment of high-temperature treatment. Since the formation of Ce^{3+} ions, as was previously shown in a number of publications [9, 19], is directly related to the formation of oxygen vacancies, the oxygen content in the atmosphere affects the content of Ce^{3+} ions in the nanocrystal. Thus, during treatment in an oxidizing medium, the content of Ce^{3+} ions in nanocrystals is negligibly small, while neutral or reducing medium stimulates the formation of oxygen vacancies and, therefore, Ce^{3+} ions.

As already was noted in Chap. 1, there is no consensus among scientists about the defect and electronic structure of cerium oxide nanocrystals, and, first of all, about the presence or absence of Ce^{3+} ions in non-stoichiometric nanocrystals as local impurity centers. While the methods of X-ray photoelectron spectroscopy and

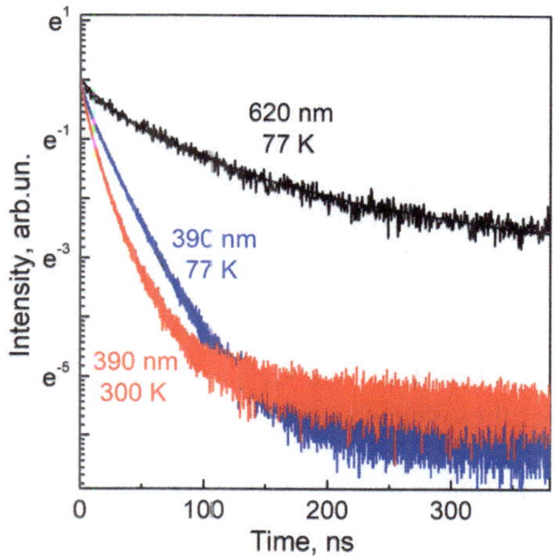

Fig. 2.2 Luminescence decay curves taken in different luminescence bands of cerium oxide nanocrystals (50 nm): 1—620 nm (77 K), 2—390 nm (293 K), 3—390 nm (77 K). (Reprinted with permission from *Journal of Luminescence*, vol. 145, A. Masalov, O. Viagin, P. Maksimchuk, V. Seminko, I. Bespalova, A. Aslanov, Yu. Malyukin, and Yu. Zorenko, Formation of luminescent centers in CeO_2 nanocrystals, pp. 61–64, 2014. Copyright 2014 Elsevier)

electron paramagnetic resonance clearly show the individual Ce^{3+} ions [20, 21] (however, these results were later criticized in [22, 23]), studies carried out using X-ray absorption spectroscopy methods [23] have shown no Ce^{3+} ions as local impurities in cerium oxide nanocrystals and delocalization of electron density between cerium and oxygen ions.

In general, the distribution of the electron density in the Ce^{4+}–O^{2-} system in cerium oxide nanocrystals remains an unresolved issue, which prevents scientists from developing adequate physical models of the electronic and defect structure of the nanocrystals. The presence in the luminescence spectra of cerium oxide nanocrystals of d-f transitions due to Ce^{3+} ions present as local impurities allows us to confirm that as oxygen is released from nanocrystals, electron density is localized precisely on cerium ions with $Ce^{4+} \rightarrow Ce^{3+}$ transitions.

2.2 Defect Centers and Luminescence Properties of Cerium Oxide Nanocrystals

2.2.1 *Ce^{3+}-Vo-Ce^{3+} Complexes*

In this section, it was shown that Ce^{3+} ions are present in non-stoichiometric cerium oxide nanocrystals (CeO_{2-x}) in the form of local impurity centers leading to 5d–4f luminescence of Ce^{3+} ions (390 nm). The formation of Ce^{3+} ions is directly related to the formation of oxygen vacancies acting as charge-compensating defects. At the same time, even among authors who agree about the presence of Ce^{3+} ions in CeO_{2-x} nanocrystals as local centers, there is a debate about the relative location of Ce^{3+} ions and related oxygen vacancies. For example, in [9], it was shown that the process of formation of oxygen vacancies in cerium oxide is significantly facilitated due to localization of two electrons remaining during release of oxygen on neighboring Ce^{4+} ions with transition to Ce^{3+} ones, which leads to the formation of Ce^{3+}-Vo-Ce^{3+} complexes (where Vo is an oxygen vacancy). Instead, other authors claim that oxygen vacancies are localized outside the first coordination sphere of the Ce^{3+} ion due to the fact that the ionic radius of the Ce^{3+} ion is significantly larger than that of the Ce^{4+} (the ionic radii of Ce^{4+} and Ce^{3+} are equal to 0.97 and 1.14 Å, respectively), and, therefore, the oxygen vacancy is repelled from the first coordination sphere [17].

So, the formation of an oxygen vacancy inside or outside the first coordination sphere of the Ce^{3+} ion changes the symmetry around cerium ion in different ways (Fig. 2.3).

As was already mentioned above, the electronic structures of both stoichiometric (CeO_2) and non-stoichiometric cerium oxide (CeO_{2-x}) were previously studied using density functional theory (DFT). It was shown that while for stoichiometric CeO_2 in the band gap, there is only an unfilled sub-band with a width of about 1 eV formed by $4f^0$ levels of Ce^{4+} ions, and for non-stoichiometric CeO_{2-x}, an additional narrow band appears in the electronic spectrum due to local $4f^1$ levels of Ce^{3+} ions at about 1.5 eV from the top of the valence band [5]. Thus, the 5d levels of Ce^{3+} ions should be located near the bottom of the conduction band of the nanocrystal.

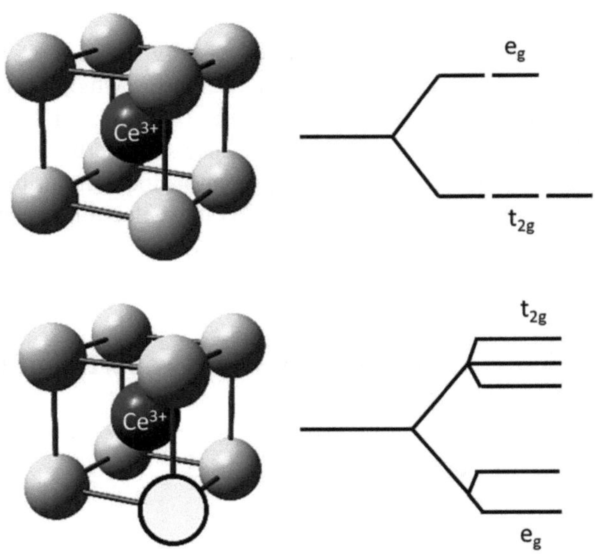

Fig. 2.3 Symmetry of the surroundings of Ce^{3+} ion and the splitting of the 5d energy level in the case of formation of oxygen vacancy outside the first coordination sphere of Ce^{3+} (the site with octahedral symmetry (O_h)) or in the first coordination sphere of the Ce^{3+} ion (the site with tetrahedral symmetry (C_{3v}))

Site symmetry of Ce^{3+} ions in CeO_{2-x} nanocrystal can be determined by the characteristic splitting of the 5d electronic states of Ce^{3+} ions, which, in turn, is determined by excitation spectra of 5d–4f luminescence. The excitation spectrum of the 390 nm luminescence band is shown in Fig. 2.4. It was found that the excitation spectrum of 390 nm luminescence band consists of five sub-bands that can be divided into two groups: the first group is two low-energy bands at 300 nm and 285 nm (4.1 eV and 4.35 eV, respectively), and the second group is three bands with higher energies at 260 nm, 246 nm, and 234 nm (4.77 eV, 5.04 eV and 5.3 eV, respectively).

Similar spectra were observed for 4f → 5d transitions of trivalent RE ions under conditions of complete removal of the 5d level degeneracy. At the same time, the nature of the splitting makes possible to determine the symmetry of the crystal environment of the rare earth ions [24–26]. Thus, if the site has an octahedral symmetry (the point group of symmetry is O_h), the 5d level splits into three low-energy and two high-energy components (Fig. 2.4). Such a situation would be observed in the non-stoichiometric CeO_{2-x} nanocrystals at non-local compensation of the excess negative charge when Ce^{4-} ion is replaced by Ce^{3+} ion. In this case, it could be argued that oxygen vacancies, which act as charge-compensating defects in CeO_{2-x}, are formed mainly outside the first coordination sphere of the cerium ion, and the symmetry of the site during the $Ce^{4+} \rightarrow Ce^{3+}$ transition remains unchanged. The experimentally observed splitting into two low-energy and three high-energy

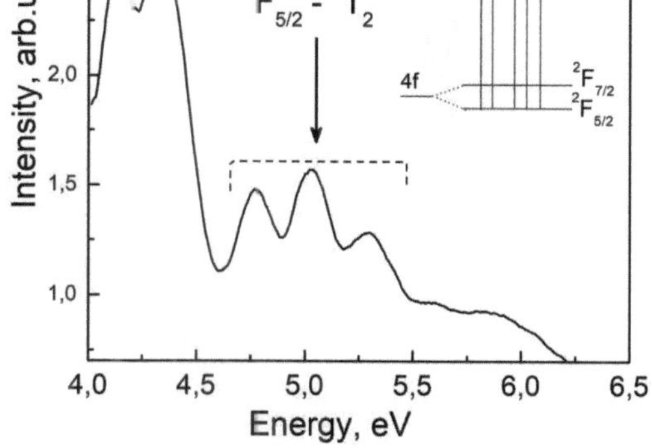

Fig. 2.4 Luminescence excitation spectrum of CeO_{2-x} nanocrystals (50 nm) after treatment in reducing atmosphere (λ_{reg} = 390 nm) obtained using synchrotron radiation at the SUPERLUMI station (Hamburg, Germany). (Reprinted with permission from *Journal of Luminescence*, vol. 145, A. Masalov, O. Viagin, P. Maksimchuk, V. Seminko, I. Bespalova, A. Aslanov, Yu. Malyukin, and Yu. Zorenko, Formation of luminescent centers in CeO_2 nanocrystals, pp. 61–64, 2014. Copyright 2014 Elsevier)

components corresponds to a lower, tetrahedral site symmetry (C_{3v}) (Fig. 2.4). A decrease in the degree of symmetry of the local center to C_{3v} can be expected if the charge-compensating defect (oxygen vacancy) is formed in the first coordination sphere of the Ce^{3+} ion.

Thus, we can say that the optical center which leads to the appearance of a luminescence band at 390 nm is Ce^{3+} ion, in the first coordination sphere of which there is an oxygen vacancy. The splitting of the luminescence band at 390 nm, observed at 77 K, can be unambiguously attributed to the splitting of the ^2F term of the Ce^{3+} ion into $^2F_{5/2}$ and $^2F_{7/2}$ ones caused by spin-orbit interaction. For oxide crystals, the value of this splitting is about 2000 cm^{-1}, which is in good agreement with the obtained experimental data (1900 cm^{-1}).

The experimentally obtained result regarding the location of oxygen vacancies in the first coordination sphere of Ce^{3+} ions agrees well with the calculations [9], according to which the formation of oxygen vacancies in cerium oxide is facilitated due to localization of two electrons remained at oxygen release, on the Ce^{4+} ions adjacent to the oxygen vacancy. As a result, Ce^{3+}-Vo-Ce^{3+} complexes (where Vo is an oxygen vacancy) are formed (Fig. 2.5). The above results indicate that such complexes are the most characteristic and widespread type of defects in the obtained nanocrystals.

Localization of electrons in the unfilled $4f^0$ levels of ions Ce^{4+} leads to the formation of filled $4f^1$ states, the energy levels of which are 1.5–2 eV below the bottom of

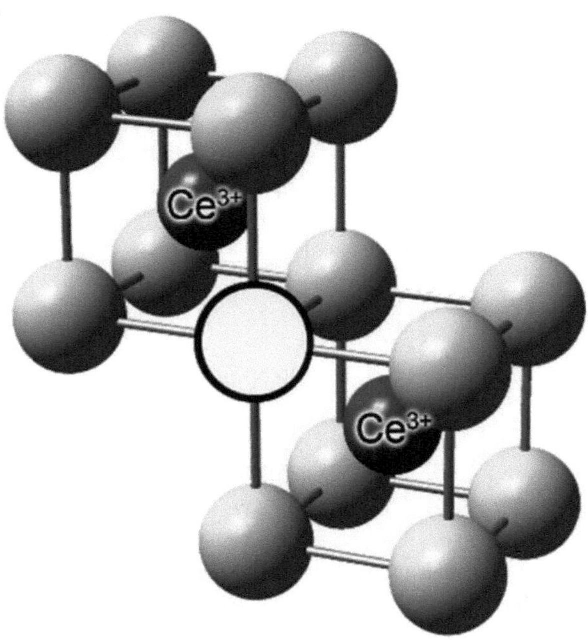

Fig. 2.5 Ce^{3+}-Vo-Ce^{3+} complexes (where Vo is an oxygen vacancy) in cerium oxide nanocrystals

the $4f^0$ sub-band. Thus, Ce^{3+} ions formed during release of oxygen from nanoparticle are local impurity centers that can manifest themselves in luminescence spectra. The authors of [9] put forward the idea that the formation of such complexes can be explained by the high content of oxygen vacancies in cerium oxide and its high oxygen capacity, as well as the ability of nanocrystals to form the C-type Ce_2O_3 structure (a structure similar to the structure of CeO_2 with 25% oxygen vacancies) in cerium oxide nanocrystals when the nanocrystal size is reduced to 2 nm or less. As will be shown in Chap. 3, the fact that Ce^{3+} ions are formed in cerium oxide nanocrystals mainly in the form of Ce^{3+}-Vo-Ce^{3+} complexes is one of the key factors that ensure the unique antioxidant activity of these nanocrystals.

Thus, in cerium oxide nanocrystals, the formation of Ce^{3+} ions is accompanied by the formation of oxygen vacancies, which together with Ce^{3+} ions form Ce^{3+}-Vo-Ce^{3+} complexes. Therefore, the intensity of luminescence (390 nm) can be considered as a measure of the content of both Ce^{3+} ions and Ce^{3+}-Vo-Ce^{3+} complexes in cerium oxide nanocrystals. This fact makes possible to observe the effect of variation of the high-temperature treatment atmosphere on the content of Ce^{3+} ions and Ce^{3+}-Vo-Ce^{3+} complexes in the studied nanocrystals. While after treatment in an oxidizing medium, the content of Ce^{3+} ions in nanocrystals is negligibly small (since the content of oxygen vacancies is small), the transition to neutral, and even more so, to reducing environment, leads to a significant increase in the content of Ce^{3+} ions, and, accordingly, the intensity of luminescence with a maximum at 390 nm.

The presence of a luminescence band caused by 5d–4f transitions of Ce^{3+} ions in cerium oxide nanocrystals makes it possible to investigate the effect of nanocrystal size on the content of Ce^{3+} ions. As was noted in Chap. 1, the unusual effect of an increase of the crystal lattice parameter of cerium oxide nanocrystals with decrease of the size was attributed in [27, 28] to the corresponding increase in the content of Ce^{3+} ions. According to the data obtained in [21] using the methods of X-ray photoelectron spectroscopy, for 30 nm cerium oxide nanocrystals, the quota of Ce^{3+} ions is 17%, for 6 nm nanocrystals, it increases to 29%, and for 3 nm nanocrystals to 44%. Electron energy loss spectroscopy studies have shown even higher values of Ce^{3+} concentration: 12.6% for 12 nm, 15.6% for 7 nm, and 53.7% for 5 nm nanoparticles [29].

X-ray photoelectron spectroscopy methods for the nanoparticles obtained in this work confirmed the results [21] regarding the high content of Ce^{3+} ions in cerium oxide nanocrystals. For example, for 2 nm nanocrystals, the content of Ce^{3+} ions, which was determined by the ratio of the areas under the Ce^{3+} and Ce^{4+} maxima in the XPS spectrum (Fig. 2.6), was about 37% (the positions of the corresponding maxima are shown in Table 2.1).

At the same time, it is known that the methods of X-ray photoelectron spectroscopy and electron energy loss spectroscopy require high-energy irradiation (X-rays or an electron beam) of experimental samples in vacuum chambers, which can lead to additional reduction of Ce^{4+} ions to Ce^{3+}. Thus, these methods can give overestimated results regarding the content of Ce^{3+} ions in nanocrystals. Optical spectroscopy methods are devoid of these drawbacks and, therefore, can give more correct estimates of the ratio of Ce^{3+}/Ce^{4+} ions in nanoparticles.

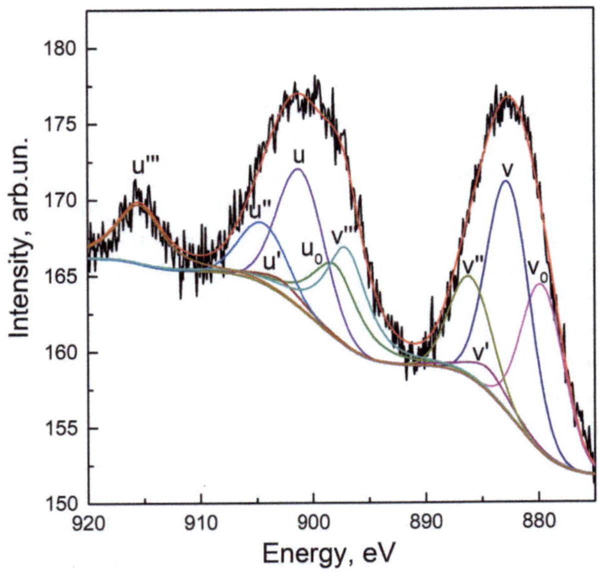

Fig. 2.6 XPS of 2 nm CeO_2 nanocrystals in the range of 870–920 eV (Ce 3d). (Reprinted with permission from *Journal of Luminescence*, vol. 242, V. Seminko, P. Maksimchuk, V. Klochkov, S. Yefimova, Luminescent colloidal ceria nanoparticles doped with RE^{3+} ions (RE = Eu, Tb) via cation exchange mechanism, pp. 118605, 2022. Copyright 2022 Elsevier)

Table 2.1 Positions of the maxima corresponding to transitions with participation of Ce^{3+} and Ce^{4+} ions in the XPS spectrum (Fig. 2.6). Reprinted with permission from *Journal of Luminescence*, vol. 242, V. Seminko, P. Maksimchuk, V. Klochkov, S. Yefimova, Luminescent colloidal ceria nanoparticles doped with RE^{3+} ions (RE = Eu, Tb) via cation exchange mechanism, pp. 118605, 2022. Copyright 2022 Elsevier

	Ce $3d_{5/2}$					Ce $3d_{3/2}$				
	v_0	v	v′	v″	v‴	u_0	u	u′	u″	u‴
E, eV	879.6	882.6	884.5	886.0	897.0	898.0	901.0	902.9	904.4	915.4
Ion	Ce^{3+}	Ce^{4+}	Ce^{3+}	Ce^{4+}	Ce^{4+}	Ce^{3+}	Ce^{4+}	Ce^{3+}	Ce^{4+}	Ce^{4+}

The results shown in Fig. 2.7 clearly indicate that reducing the size of cerium oxide nanocrystals from 50 nm to 2 nm really leads to a significant increase in the content of Ce^{3+} ions in nanocrystals (approximately, in ten times), which is evident from the corresponding change in the intensity of Ce^{3+} luminescence and correlates well with the results obtained by other authors [28, 30]. When the size of the nanocrystals decreases, the maximum of the Ce^{3+} luminescence band shifts to longer wavelengths (390 nm for 50 nm nanoparticles, 410 nm for 10 nm nanoparticles, and 430 nm for 2 nm nanoparticles). This shift can be explained by the growth of the unit cell parameter, which, as was shown in a number of publications [27–30], is observed when the size of nanocrystals decreases, and according to [32] can reach up to 7% (from 5.41 Å for bulk samples up to 5.61 Å for 1–2 nm nanoparticles).

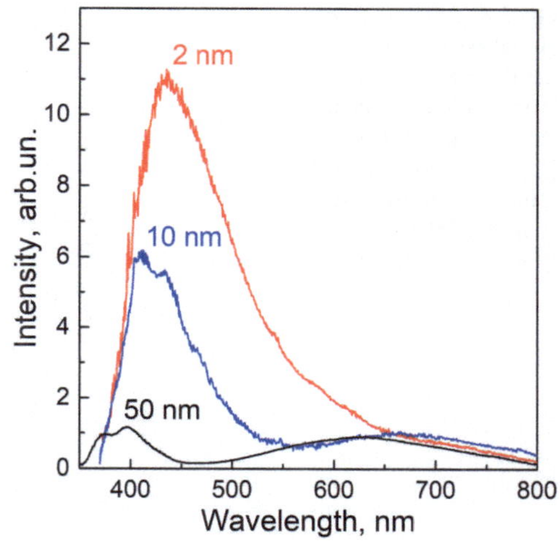

Fig. 2.7 Luminescence spectra of CeO_2 nanocrystals of different sizes (2 nm, 10 nm, 50 nm)

As was already mentioned, today the most controversial issues regarding the defect and electronic structure of cerium oxide nanocrystals are: (1) the presence or absence of Ce^{3+} ions in cerium oxide nanocrystals as local impurity centers; (2) relative location of Ce^{3+} ions and oxygen vacancies in nanocrystals; (3) dependence of the content of Ce^{3+} ions (and, therefore, oxygen vacancies) on the size of the nanocrystal. The use of luminescence spectroscopy methods to study the presence, relative location, and content of Ce^{3+} ions in cerium oxide nanocrystals of various sizes allowed us to confirm that Ce^{3+} ions are present in cerium oxide nanocrystals as local impurity centers, form complexes with oxygen vacancies (Ce^{3+}-Vo-Ce^{3+}), and their content increases both during transition from an oxidizing to a reducing atmosphere and when the nanocrystal size decreases.

2.2.2 *F⁰-Centers and Charge-Transfer Complexes*

In Sect. 2.2.1, the nature of the band with a maximum at 390 nm in the luminescence spectra of cerium oxide nanocrystals was determined. In this subsection, we will consider the band with a maximum at 620 nm, which is also present in the spectra (Fig. 2.1).

The excitation spectrum of the band with a maximum at 620 nm is shown in Fig. 2.8. Unlike the excitation spectrum of the band at 390 nm, this spectrum consists of one broad band (~4000 cm^{-1}) with a maximum at 340 nm.

A similar band was previously observed in the excitation spectra of cerium oxide nanocrystals doped by Eu^{3+} and Sm^{3+} ions and was attributed to the Ce^{4+}-O^{2-} charge-transfer transitions [33]. It was previously shown [34] that in the absorption spectra of cerium-containing oxide crystals (in particular, Sr_2CeO_4), transitions with charge transfer (CT transitions) from the oxygen ion (O^{2-}) to the cerium ion (Ce^{4+}) are

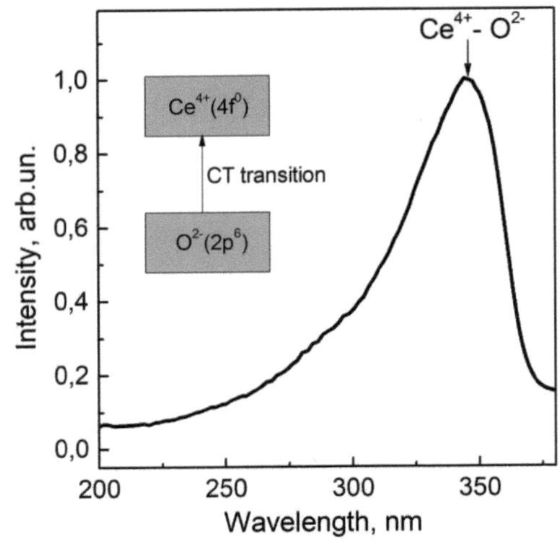

Fig. 2.8 Luminescence excitation spectrum of CeO_{2-x} nanocrystals (50 nm) after treatment in reducing atmosphere ($\lambda_{reg} = 620$ nm) obtained using synchrotron radiation at the SUPERLUMI station (Hamburg, Germany)

observed, which in some cases can relax in the form of CT-luminescence. CT-luminescence is usually observed only at low temperatures, while non-radiative relaxation occurs at room temperature.

In [35], it was shown that the excitation spectra of both doped and undoped CeO_{2-x} nanocrystals contain a broad band with a maximum at 340 nm, which corresponds to charge-transfer transition from oxygen to cerium ion. The position of the band corresponding to the charge-transfer transition in the excitation spectrum can be roughly estimated by Jorgensen's formula [36]:

$$\sigma\left(\text{cm}^{-1}\right) \approx 30{,}000\left(\text{cm}^{-1}\right) \bullet \left[\chi_{\text{opt}}\left(X\right) - \chi_{\text{opt}}\left(M\right)\right], \tag{2.1}$$

where $\chi_{opt}(X)$ and $\chi_{opt}(M)$ are optical electronegativities of the ligand and the metal ion, respectively.

For oxygen and cerium ions χ_{opt} $(O^{2-}) = 3.2$ eV and $\chi_{opt}(Ce^{4+}) = 2.1$ eV, so, $\sigma \approx 33{,}000$ cm^{-1} and $\lambda \approx 310$ nm, which is close to the experimentally obtained value (340 nm). For Sr_2CeO_4, charge-transfer band was also observed in this range ($\lambda_{max} = 335$ nm) [34].

A more detailed analysis of the band with a maximum at 620 nm shows that it consists of two bands, the intensities of which change in different ways with variation of temperature and excitation wavelength. For instance, temperature increase from 77 to 300 K leads to the shift of the peak observed at 325 nm excitation from 620 nm to 570 nm, which indicates the inhomogeneous nature of the 620 nm band (Fig. 2.9).

Excitation with $\lambda_{exc} = 325$ nm (3.8 eV) corresponds to the charge-transfer transition ($O2p^6 \rightarrow Ce4f^0$) in the cerium oxide lattice, and the reverse relaxation of the excited state can occur either due to reverse charge-transfer (CT-luminescence) or via luminescence of defect states (such as F centers). Authors [13, 37] discussed the possibility of both defect and CT-luminescence for cerium oxide nanocrystals. In order to clarify the nature of the luminescence band with a maximum at 620 nm,

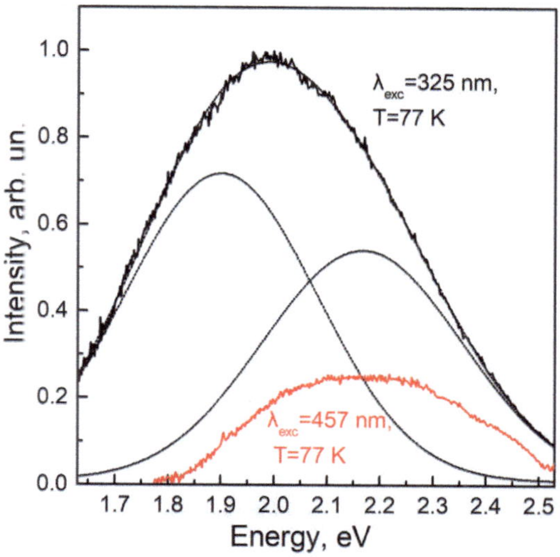

Fig. 2.9 Luminescence spectra of CeO_2 nanocrystals treated in an oxidizing medium at $\lambda_{exc} = 325$ nm and 457 nm at different temperatures (77 K and 300 K) and fitting the luminescence spectrum at $\lambda_{exc} = 325$ nm using Gaussian functions

the luminescence was excited at different wavelengths (457 nm, 488 nm, and 532 nm).

At λ_{exc} = 457 nm (77 K), the luminescence spectrum consists of a band with maximum at 570 nm (2.17 eV), which coincides with the luminescence band at λ_{exc} = 325 nm (300 K), and the same luminescence spectra are observed at λ_{exc} = 532 nm and 488 nm. The luminescence band obtained at excitation with λ_{exc} = 325 nm (77 K) can be fitted by two Gaussian functions with maxima at 1.9 eV (650 nm) and 2.17 eV (570 nm), the position of the latter corresponds to the maximum of the band observed at λ_{exc} = 457 nm. Thus, at λ_{exc} = 325 nm, one can excite both defect (570 nm) and CT-luminescence (650 nm), while at λ_{exc} = 457 nm, only the luminescence related to transitions between defect levels in the band gap is observed. The excitation spectra obtained at registration in both luminescence bands (570 nm and 650 nm) are shown in Fig. 2.10.

Both spectra have a band corresponding to charge-transfer transition (~3.4 eV), but the spectrum obtained at λ_{reg} = 570 nm also contains a band with a maximum at 470 nm (2.65 eV). In contrast to the luminescence band with a maximum at 570 nm, the luminescence band with a maximum at 650 nm was not observed under any sub-band gap excitation and therefore can be related to radiative relaxation in the excited charge-transfer complex Ce^{4+}-O^{2-} (CT-luminescence). The high value of the Stokes shift (more than 13,000 cm^{-1}) and the strong temperature quenching of this band confirm this assumption.

As was already mentioned, the processes of temperature quenching for the defect band (570 nm) and the charge-transfer band (650 nm) differ sufficiently. Figure 2.11 shows that stronger temperature quenching of CT-luminescence leads to the dominance of the defect luminescence band in the spectra at high temperatures (more than 180 K). By studying the change in the intensity of the luminescence bands at temperature variation, the activation energies of temperature quenching for both bands were determined.

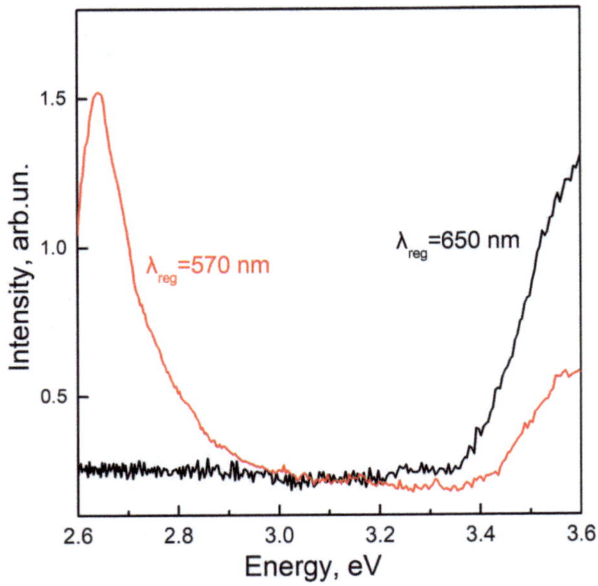

Fig. 2.10 Excitation spectra of 570 and 650 nm luminescence bands in CeO_2 nanocrystals treated in oxidizing medium at $T = 77$ K

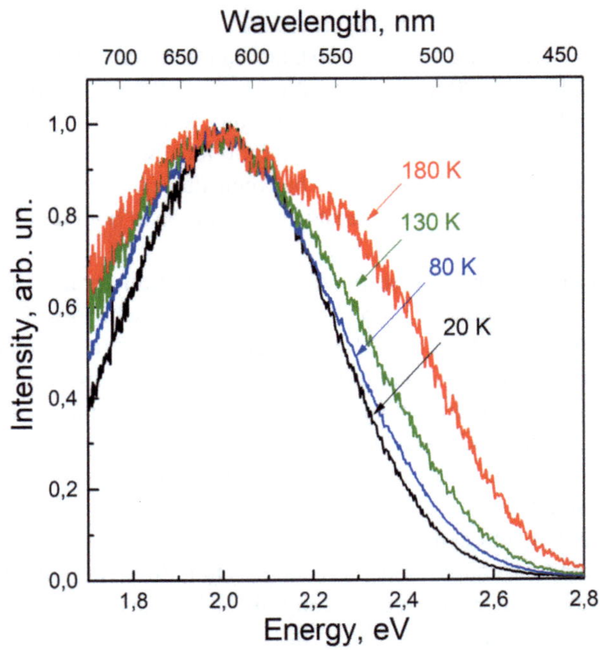

Fig. 2.11 Redistribution of the luminescence intensities of the defect band (570 nm) and charge-transfer band (650 nm) in cerium oxide nanocrystals at different temperatures (1—180 K, 2—130 K, 3—80 K, 4—20 K). The spectra are normalized to intensity of the charge-transfer band (650 nm)

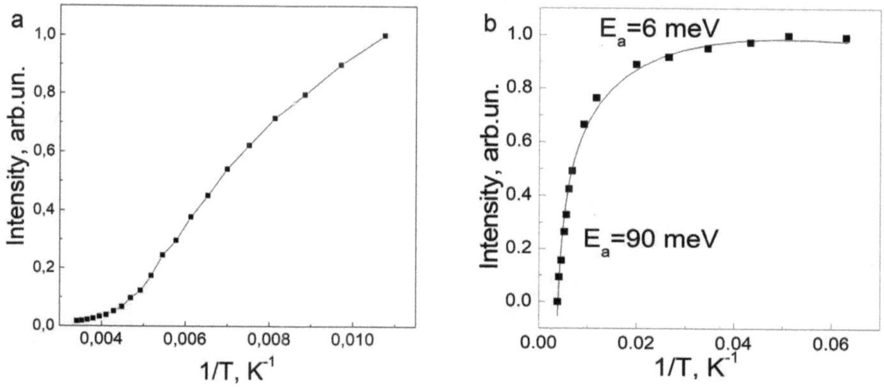

Fig. 2.12 Temperature dependences of charge-transfer band (650 nm) (**a**) and defect band (570 nm) (**b**) intensities

The temperature dependence of the intensities of both bands can be described by:

$$I = I_0\left(1 + B_1 e_1^{-E_1/kT} + B_2 e_2^{-E_2/kT}\right), \tag{2.2}$$

where E_1 and E_2 are trap energies of the first and second type, respectively [38].

The fitting makes it possible to determine the activation energies of temperature quenching at low and high temperatures. For charge-transfer band (650 nm), the activation energy is 5 meV at low and 30 meV at high temperatures, and for defect band (570 nm), these values are 6 meV and 90 meV, respectively (Fig. 2.12).

The value of the activation energy of temperature quenching of 570 nm band is in good agreement with the value of the activation energy for F^0 centers in cerium oxide nanocrystals obtained by the coprecipitation method [13]. In addition, the maximum of the luminescence band formed by F^0 centers determined in [13] (2.1 eV) is close to the maximum of the defect luminescence band for the investigated nanocrystals (2.17 eV). Thus, the luminescence band with a maximum of 570 nm is most likely formed by transitions of F^0 centers ($F^{0*} \rightarrow F^0$). At the same time, the excited levels of the F^0 centers (F^{0*}) should be approximately 0.1 eV below the bottom of the $4f^0$ sub-band, which causes a strong temperature quenching of this band (which, however, is weaker than for the charge-transfer band).

Therefore, upon excitation with charge transfer from the valence band to the $4f^0$ sub-band of cerium oxide nanocrystals, energy relaxation can occur either by charge-transfer luminescence or by electron capture by oxygen vacancies and radiative relaxation of the resulting F^0 centers. Studies of the energy relaxation processes at registration in the bands of charge-transfer luminescence (650 nm) and F^0 centers (570 nm) show that the dynamics of electronic excitations for these transitions is significantly different.

Figure 2.13 shows the decay curves of charge-transfer luminescence bands and F^0-centers (a and b, respectively) upon excitation in the charge-transfer band (325 nm).

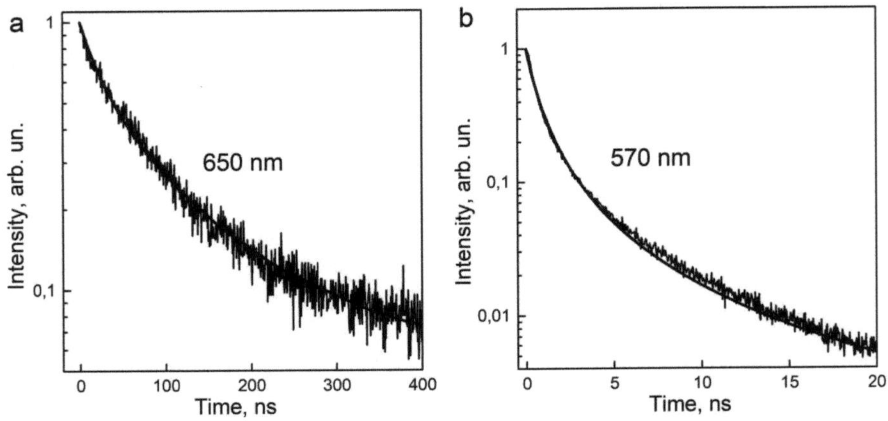

Fig. 2.13 Luminescence decay curves of the charge-transfer band (650 nm) (**a**) and the defect band (570 nm) (**b**) for cerium oxide nanocrystals treated in an oxidizing medium, $\lambda_{exc} = 325$ nm

Both decay curves are non-exponential and can be approximated by the hyperbolic Becquerel's law [39]:

$$I = I_0 \left(1 + \mathrm{pt}\right)^{-\alpha}, \tag{2.3}$$

where p and α are empirical parameters.

However, the approximation parameters p and α for these bands are quite different, for 570 nm band: $p = 0.83$ and $\alpha = 1.83$, and for 650 nm band: $p = 0.02$ and $\alpha = 1.17$. It can also be seen that while the decay time of the defect band (570 nm) is few nanoseconds, the decay time of the charge-transfer luminescence (650 nm) is much longer and is about 150 ns. These results cannot be explained by the different gaps between these levels and $4f^0$ sub-band, since, as it was shown from the analysis of the temperature quenching of both luminescence bands, the activation energy for the band of F^0 centers is higher than the corresponding energy for the charge-transfer luminescence band (90 meV and 30 meV, respectively), so one would expect the charge-transfer luminescence lifetime to be shorter.

It can be assumed that the long lifetime of charge-transfer excitation is due to its high mobility. Indeed, as was shown in [39], this type of decay curves is observed when the probability of excitation capture and re-release is significant. According to [40], the value of the parameter $\alpha = 1.17$, which was observed for charge-transfer luminescence, corresponds to ≈ 25 acts of capture and re-release per act of emission, while at $\alpha = 1.83$ (the luminescence band F^0-centers), the probability of capture and re-release should be approximately equal.

Thus, the charge-transfer excitation is characterized by a high recapture probability, while the recapture probability of the excitation corresponding to the 570 nm band is significantly lower. Several acts of capture and re-release should be considered as an evidence for the presence of small traps near the bottom of the $4f^0$ sub-band. F^0-centers cannot act as such traps, since in this case there should be an increase in the luminescence intensity at 570 nm at the initial stage of the decay curve (due to energy transfer from charge-transfer complexes). Since this effect is not observed, the electron traps must form an independent subsystem. Methods of thermoluminescence analysis were used to investigate the presence of electronic traps, their depth, and the nature of these traps.

2.2.3 Electron Traps

The significant number of charge-transfer excitation recapture events shown in the previous subsection may be due to the presence of small electron traps below the bottom of the $4f^0$ sub-band. The study of the subsystem of electron traps was carried out using thermostimulated luminescence (TSL) methods. The TSL technique is based on the fact that phosphor initially excited in order to fill the traps with charge carriers, can show luminescence during heating. Keeping the heating rate constant, a curve of thermoluminescence intensity temperature dependence, so-called thermoluminescence curve, is obtained. The maxima on this curve correspond to certain traps from which electrons are released at the highest rate at the temperature corresponding to the thermoluminescence maximum.

TSL curves were studied to determine the depth of electron traps of cerium oxide nanocrystals treated in reducing and oxidizing media after irradiation with He-Cd laser (325 nm) for 15 min. TSL curves are shown in Fig. 2.14. As can be seen from the figure, there is only one peak on the TSL curve at 120 K regardless of the oxygen stoichiometry. The position of this maximum coincides for nanocrystals treated in different environments, and only the integral intensity of the maximum differs. Based on this, it can be concluded that the nature of the defects that determine the appearance of the thermoluminescence peak is similar for cerium oxide

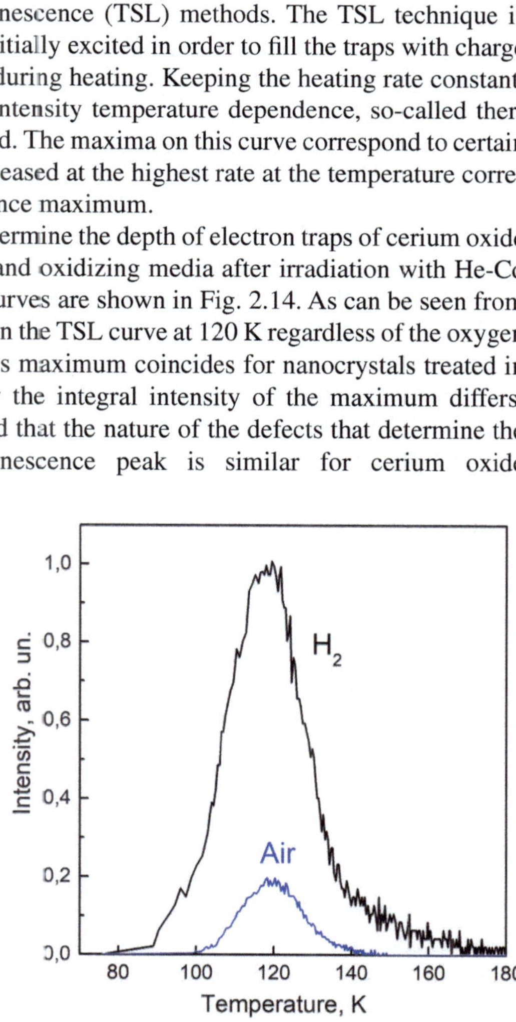

Fig. 2.14 TSL curves of CeO_{2-x} nanocrystals treated in reducing (H_2) and oxidizing (Air) environments. TSL was measured in the Ce^{4+}-O^{2-} charge-transfer band (650 nm)

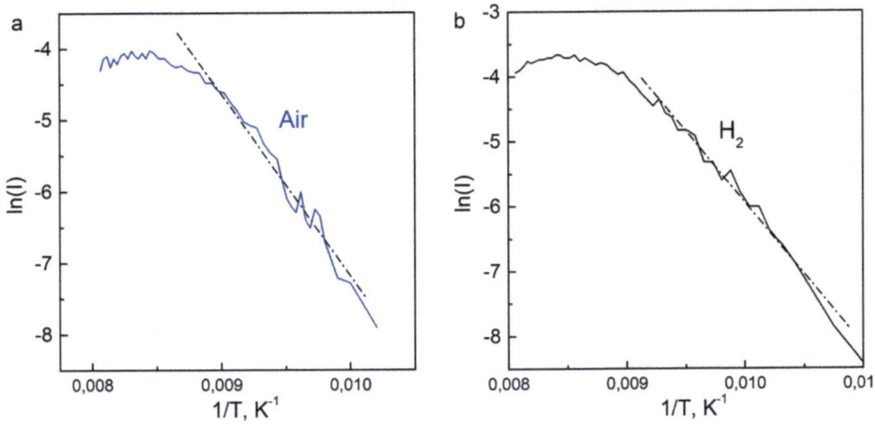

Fig. 2.15 Calculation of the activation energy of electron traps by the method of analysis of the initial areas of thermoluminescence maxima for cerium oxide nanocrystals treated in oxidizing (**a**) and reducing (**b**) environments

nanocrystals treated in reducing and oxidizing environments. The increase in the integral intensity of the thermoluminescence band with increasing oxygen deficiency in CeO_2 nanocrystals allows us to conclude that the band on the thermoluminescence curve is related to oxygen vacancies.

The energy of activation of electronic traps was determined by the method of analysis of initial part of thermoluminescence peak. According to this method, the initial part of the rise of the thermoluminescence curve depends exponentially on the temperature. Thus, the region of the initial growth of the curve in the coordinates ln (I) to 1/T should be a straight line with the slope: $-E_a/k$, from which the activation energy E_a can be easily found (Fig. 2.15).

The calculated value of the trap depth was 0.21 eV for nanocrystals treated in an oxidizing environment and 0.19 eV for nanocrystals treated in reducing environment. The close values of the activation energy of electron traps confirm the assumption that the nature of these traps for cerium oxide nanocrystals treated in reducing and oxidizing environments is similar.

TSL activation energy (200 meV) is higher than the activation energy of temperature quenching of the 570 nm band (90 meV), therefore, as mentioned above, electron traps cannot be related to F^0 centers. However, the values of the activation energy of this peak is quite close to the position of F^+-center level calculated in [13], which makes it possible to associate these electronic traps with F^+ centers.

Therefore, the transition of a large number of Ce^{4+} ions to Ce^{3+} is associated with the oxygen removal processes from the surface of cerium oxide nanocrystals during treatment in a reducing environment. The removal of each oxygen ion is accompanied by the formation of oxygen vacancies that can capture electrons. During treatment in reducing environment, the number of oxygen vacancies increases and their presence leads to a significant modification of the excitation energy relaxation processes. As a result of the capture of electrons from the $4f^0$ sub-band by oxygen vacancies, F^0 and F^+ centers are formed, which can play the role of both luminescent centers (which was shown in Sect. 2.2.2) and electron traps (which was shown in this subsection).

The intensity of the TSL peak of cerium oxide nanocrystals treated in a reducing medium is ~5 times greater than that of cerium oxide nanocrystals treated in an oxidizing medium due to much higher content of oxygen vacancies in cerium oxide nanocrystals treated in reducing medium. However, as can be seen from the luminescence spectra and TSL curves, even after treatment in oxidizing medium, there is a certain concentration of oxygen vacancies in the nanocrystals, which can form F^0 and F^+ centers. This may be due, in particular, to the relatively low energy of oxygen vacancy formation in cerium oxide nanocrystals.

2.2.4 Energy Levels of Stoichiometric and Non-stoichiometric Cerium Oxide Nanocrystals

Summarizing the results of the experimental studies described in this chapter, it can be concluded that the transition from stoichiometric to non-stoichiometric cerium oxide nanocrystals leads to significant change in both defect structure and electronic properties of nanocrystals directly affecting the luminescent characteristics of nanocrystals. The removal of oxygen from the surface of the nanocrystal during treatment in reducing medium is accompanied by formation of oxygen vacancies and the capture of electrons remained during the release of oxygen either by Ce^{4+} ions or by the oxygen vacancies leading to formation of a number of new bands in the luminescence spectra of cerium oxide nanocrystals.

The scheme of levels of stoichiometric cerium oxide is well known. Cerium oxide (CeO_2) has a band gap of about 6 eV [5]. A narrow (1 eV) sub-band formed by $4f^0$ states of Ce^{4+} ions is located 3.15 eV above the top of the valence band.

The generalization of the experimentally obtained results made it possible to construct a general scheme of energy levels for CeO_{2-x} nanocrystals (Fig. 2.16). The analysis of the processes of temperature quenching of the luminescence band of F^0 centers shows that their excited levels are 0.1 eV below the $4f^0$ sub-band, while the analysis of TSL curves make it possible to determine that the levels of F^+ centers, which form electron traps for charge-transfer excitations, are 0.2 eV below the $4f^0$ sub-band leading to a significant number of acts of excitation recapture during the lifetime.

After treatment in reducing environment, the degree of oxygen non-stoichiometry of nanocrystals increases significantly, which leads to formation of new bands in the luminescence spectra.

Previously, in theoretical and experimental studies by various authors, it was shown that for non-stoichiometric cerium oxide nanocrystals, the transition of some ions from Ce^{4+} to Ce^{3+} leads to the splitting of the $4f^0$ band into two sub-bands—filled ($4f^1$) and empty ($4f^0$) ones [5, 41, 42]. According to various authors, the gap

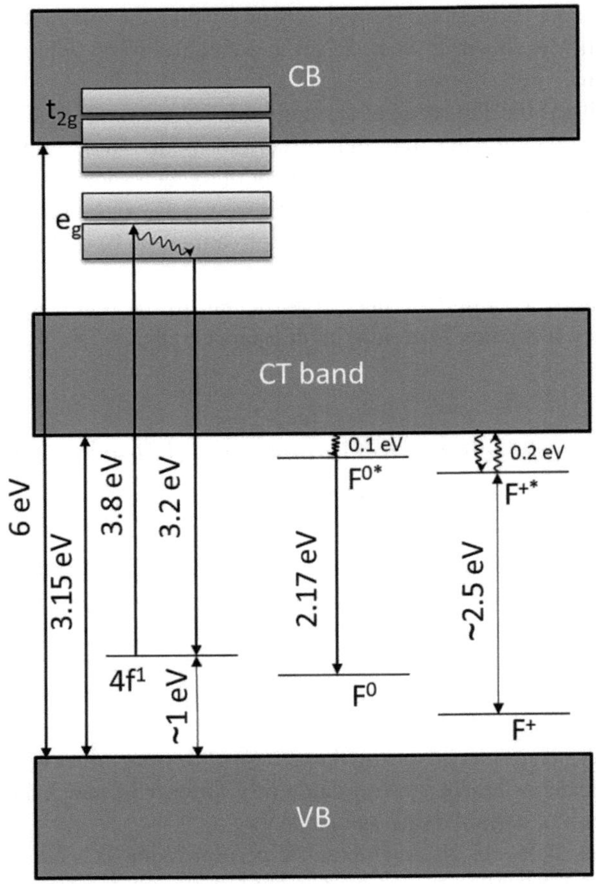

Fig. 2.16 Scheme of energy levels for cerium oxide nanocrystals after treatment in reducing environment (providing high degree of oxygen non-stoichiometry), positions of 4f and 5d levels of Ce^{3+} ions in the band gap and levels of defects associated with F centers are shown. The $4f^1$ level position was taken from [5], the position of the ground level of F^+-centers was taken from [13]

between the valence band and the $4f^1$ band is about 0.7–1 eV [41] or 1.2–1.5 eV [5, 42]. This makes it possible to use the experimental data obtained in this work to determine the position of the 5d levels on the energy level diagram.

The analysis of luminescence spectra and luminescence excitation spectra, as well as decay curves, given in Sect. 2.1, shows that the luminescence band with a maximum at 390 nm for cerium oxide nanocrystals treated in a reducing medium is caused by 5d → 4f luminescence of Ce^{3+} ions. In turn, from the excitation spectra of this band, the relative position of the 5d energy levels of Ce^{3+} ions on the general scheme of energy levels can be determined. If we assume that the data given in [5] are correct, and taking into account the data obtained from the excitation spectra, the lowest $5d^1$ component should be 3.8 eV above the main $4f^1$ level, therefore the lowest $5d^1$ component of Ce^{3+} ions is at ~1 eV below the bottom of the conduction band. A similar position of the excited 5d states relative to the conduction band should lead to a noticeable temperature quenching of Ce^{3+} luminescence and may be one of the reasons why the intensity of Ce^{3+} ion luminescence is relatively low compared to other crystal lattices doped by Ce^{3+} ions.

2.3 The Influence of Impurity Ions on the Formation of the Defect Structure of CeO_2-ZrO_2, CeO_2-Y_2O_3, CeO_2-Tb_2O_3, CeO_2-Eu_2O_3 Nanocrystals

2.3.1 The Influence of Aliovalent Ions on the Processes of Formation of Luminescent Centers in Cerium Oxide Nanocrystals

The presence of trivalent impurity ions in cerium oxide leads to formation of additional oxygen vacancies necessary to compensate the excess negative charge. The relative positions of oxygen vacancies and rare earth ions in RE-doped cerium oxide nanocrystals were studied recently in several works [14, 43]. In [17], it was shown that for ions with radius smaller than the radius of Gd^{3+} ion, the oxygen vacancies are located mainly in the first coordination sphere of the rare earth ion, while for ions with radius larger than the radius of the Gd^{3+} ion (such as Eu^{3+} ions), the oxygen vacancies are located outside of the first coordination sphere of the rare earth ion. Here the relative location of oxygen vacancies and impurity ions in cerium oxide nanocrystals was investigated using optical spectroscopy methods.

The effect of aliovalent ions on the defect structure and luminescent properties was considered for nanocrystals of mixed cerium-yttrium oxides (CeO_2-Y_2O_3). Cerium-yttrium oxide is known as an ionic conductor, which is widely used in solid oxide fuel cells (solid oxide fuel cells) as an electrolyte. Data on the optimal concentrations of yttrium in this material differ sufficiently, for instance, in [44] was shown that the replacement of a part of Ce^{4+} ions by Y^{3+} ions leads to an increase in the electrical conductivity of the material only at low concentrations of impurity ions (up to 1 at.%), while the authors [45] claimed that the optimal concentrations were ~20 at.%. The dependence of the ratio of the intensities of Ce^{3+} 5d → 4f and F^0 bands on the content of oxygen vacancies opens the possibility of studying the processes of the formation of oxygen vacancies in CeO_2-Y_2O_3 nanocrystals using the methods of luminescence spectroscopy.

All bands that are observed in the luminescence spectrum of cerium oxide (5d → 4f luminescence of Ce^{3+} ions, transitions of F^0 centers, and CT-luminescence) were also observed in the spectra of mixed oxides, but the ratio of their intensities for mixed cerium-yttrium oxide is different, which indicates the influence of impurity ions on the contents of optical centers of different types. Figure 2.17 shows the luminescence spectra of CeO_2-Y_2O_3 nanocrystals with concentrations of yttrium ions equal to 1, 3, and 10 at.% at 10 K after high-temperature treatment in vacuum.

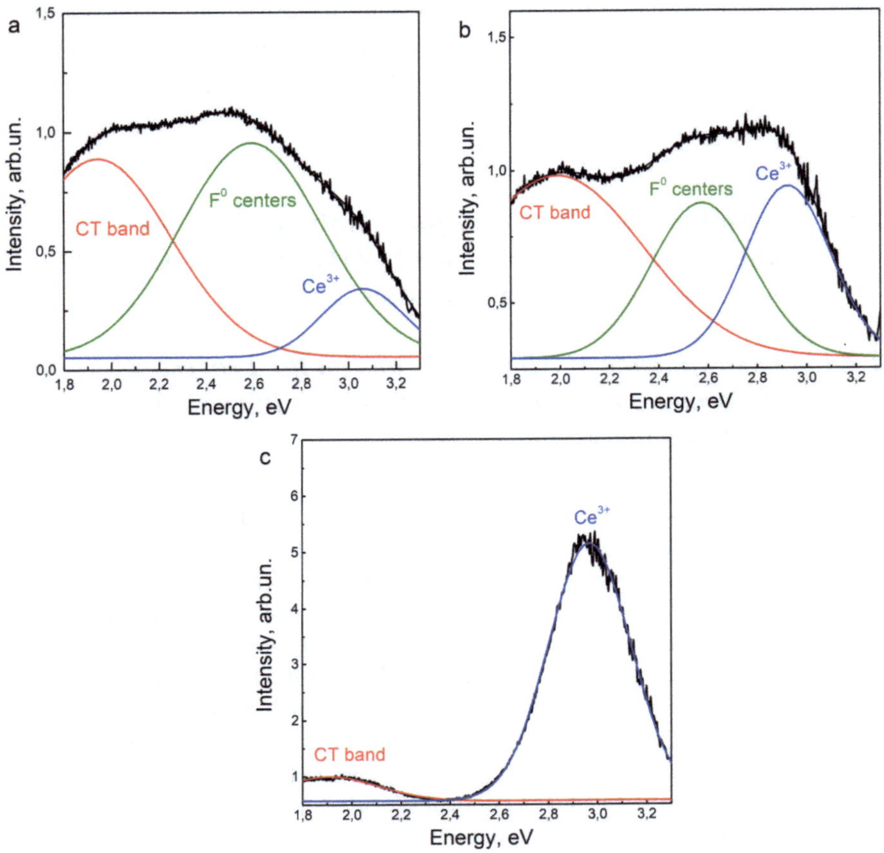

Fig. 2.17 Luminescence spectra of CeO_2-Y_2O_3 nanocrystals (T = 10 K) with different yttrium content: (**a**) 1 at.%; (**b**) 3 at.%; (**c**) 10 at.%

The luminescence spectrum of mixed CeO_2-Y_2O_3 nanocrystals consists of three bands with maxima at 420 nm (3 eV), 480 nm (2.6 eV), and 630 nm (1.97 eV), which are caused by 5d → 4f transitions of Ce^{3+} ions, F^0-centers, and CT-luminescence, respectively. With an increase in the concentration of yttrium ions, there is an increase in the intensity of the luminescence band caused by 5d → 4f transitions of Ce^{3+} ions. At concentration of yttrium ions of 1 at.%, the intensity of the band related to F^0-centers is the highest, at further concentration increase its intensity decreases. At concentration of yttrium ions of about 10 at.%, the band related to F^0 centers is not observed. At the same time, the 420 nm band caused by 5d → 4f transitions of Ce^{3+} ions, grows monotonically in the entire range of yttrium concentrations from 1 to 10 at.%.

When Ce^{4+} ions are replaced by aliovalent Y^{3+} ions, charge-compensating defects (oxygen vacancies) are formed. Since oxygen vacancies are necessary both for the formation of F^0 centers (Vo + $2e^-$) and for the compensation of the excess charge of

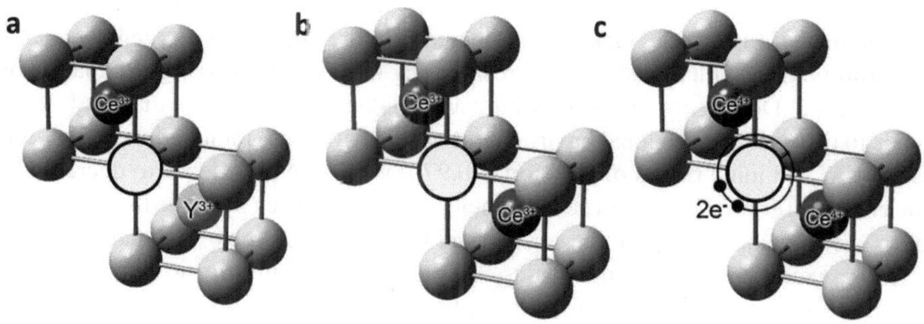

Fig. 2.18 The main types of defect centers in CeO_2-Y_2O_3 nanocrystals: (**a**) Ce^{3+}-Vo-Y^{3+} complexes; (**b**) Ce^{3+}-Vo-Ce^{3+} complexes; (**c**) F^0-centers

Ce^{3+} ions by binding vacancies into Ce^{3+}-Vo-Ce^{3+} and Ce^{3+}-Vo-Y^{3+} complexes (Fig. 2.18), the dependence of the content of F^0-centers on the concentration of Y^{3+} ions can be explained by the fact that at low concentrations, (about 1 at.%) oxygen vacancies preferentially capture electrons and form F^0-centers, while at high concentrations, the main type of defects are Ce^{3+}-Vo-Y^{3+} and Y^{3+}-Vo-Y^{3+}complexes.

Since at high concentrations of yttrium ions, the content of Ce^{3+}-Vo-Ce^{3+} complexes is significantly lower than that of Ce^{3+}-Vo-Y^{3+} complexes, the concentration of Ce^{3+} ions in nanocrystals is proportional to the concentration of Y^{3+} ions reaching up to 10 at.% at the corresponding concentration of Y^{3+} ions explaining the significant increase in the intensity of Ce^{3+} luminescence. A similar increase in the intensity of Ce^{3+} luminescence with an increase in the concentration of impurity ions was also observed for other aliovalent ions (Tb^{3+}, Eu^{3+}) due to the fact that at high concentrations of aliovalent impurities, the most common type of defects was Ce^{3+}-Vo-Re^{3+} complexes (where Re = Tb, Eu).

2.3.2 *The Influence of Isovalent Ions on the Processes of Formation of Luminescent Centers in Cerium Oxide Nanocrystals*

The influence of isovalent ions on the defect structure and luminescent properties of cerium oxide nanocrystals was considered for cerium-zirconium (CeO_2-ZrO_2) oxide nanocrystals. Mixed cerium-zirconium oxides (CeO_2-ZrO_2) attract a lot of attention as promising materials for the development of oxygen storage devices when used together with metal catalysts [46, 47]. The high oxygen capacity of cerium-zirconium oxides allows them to accumulate and release high concentrations of oxygen, which makes it possible to reduce fluctuations in the oxygen content in environment during cycling of the metal catalyst.

Cerium-zirconium oxides, in particular, are widely used in automotive converters, which allow the simultaneous removal of carbon monoxide, nitrogen oxide, and unburned hydrocarbons [48]. It was previously shown that replacing a part of Ce^{4+} ions with Zr^{4+} (up to 40%) leads to an increase in the content of oxygen vacancies in cerium oxide, which leads to a corresponding increase in the oxygen capacity of the material. Doping by Zr^{4+} ions, which are smaller in ionic radius than Ce^{4+} ions, leads to distortion of the local environment of impurity ions (ionic radius of Zr^{4+} ion is 0.84 Å, and ionic radius of Ce^{4+} ion is 0.97 Å). This, in turn, leads to decrease in the energy of formation of oxygen vacancies, and thus makes possible the formation of oxygen vacancies both on the surface and within the volume of the material [49].

For mixed cerium-hafnium oxides, the influence of impurity ions on the processes of oxygen vacancy formation and oxygen accumulation/release has been studied much less than for cerium-zirconium oxides (in particular, in [50]), although, based on the similarity of the ionic radii of Zr^{4+} ions (0.84 Å) and Hf^{4+} (0.83 Å), the effect of hafnium ions should be similar to the effect of zirconium ions.

As in the case of mixed cerium-yttrium oxides, presence of Zr^{4+} ions in the lattice of cerium oxide leads to a noticeable increase in the concentration of oxygen vacancies and, accordingly, of both types of optical centers formed with involvement of oxygen vacancies (Ce^{3+}–Vo–Ce^{3+} complexes and F^0-centers). In contrast to CeO_2-Y_2O_3 nanocrystals, the presence of Zr^{4+} ions does not lead directly to the formation of the corresponding concentrations of oxygen vacancies, but only to a decrease in the energy of their formation, so the content of oxygen vacancies should depend not only on the concentration of impurity ions, but also on the environment of high-temperature treatment.

Figure 2.19a, b shows the luminescence spectra of nanocrystals of mixed cerium-zirconium oxides treated in an oxidizing environment with concentrations of zirconium ions of 10 at.% and 20 at.% at $T = 10$ K. Luminescence spectrum of mixed oxides treated in oxidizing environment consists of the band with maximum at 410 nm (3 eV) related to 5d → 4f transitions of Ce^{3+} ions, and the band with maximum at 630 nm (1.95 eV), which consists of two components with maxima at

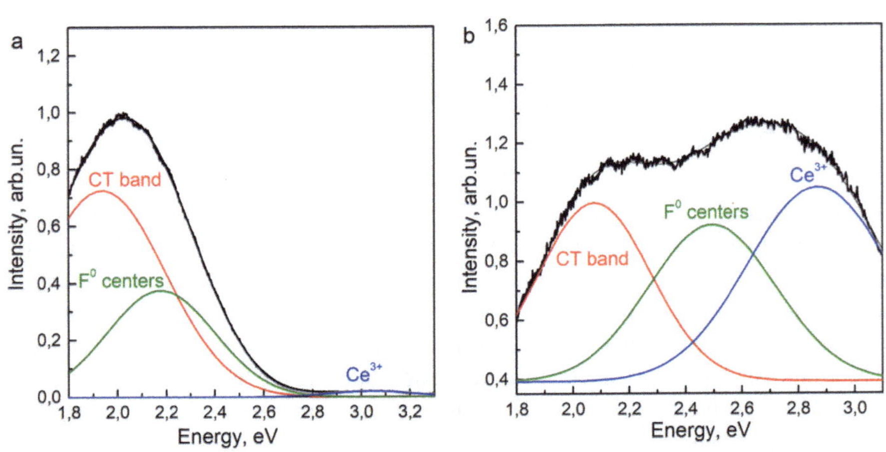

Fig. 2.19 Luminescence spectra of CeO_2-ZrO_2 nanocrystals ($T = 10$ K) with different zirconium content: (**a**) 10 at.%; (**b**) 20 at.%, after treatment in an oxidizing environment

570 nm (2.17 eV) and 640 nm (1.9 eV) related to F^0 centers and charge-transfer luminescence, respectively. With an increase in the concentration of zirconium ions, there is a blueshift of the band of F^0 centers and redshift of the band of Ce^{3+} ions.

Unlike cerium oxide, for mixed cerium-zirconium oxides, the band related to 5d → 4f transitions of Ce^{3+} ions is observed even after treatment in oxidizing environment, and its intensity increases with increasing concentration of zirconium ions (Zr^{4+}). So, in CeO_2-ZrO_2 nanocrystals, even after treatment in oxidizing environment, the content of oxygen vacancies is quite large due to the low energy of formation of oxygen vacancies in mixed nanocrystals.

The luminescence spectrum of mixed CeO_2-ZrO_2 oxides treated in reducing environment (Fig. 2.20a, b) consists of three bands with maxima at 440 nm (2.8 eV), 500 nm (2.5 eV), and 610 nm (2.05 eV), which are caused by 5d → 4f transitions of Ce^{3+} ions, F^0 centers, and charge-transfer luminescence, respectively. With an increase in the concentration of zirconium ions, as in the case of treatment in oxidizing environment, a blueshift of the band of F^0 centers is observed. The presence of 10 at.% of zirconium ions leads to a threefold, and 20 at.% to a sixfold intensity increase of the band caused by luminescence of Ce^{3+} ions, and, accordingly, to an equally high increase in the concentration of Ce^{3+}-Vo-Ce^{3+} complexes.

In contrast to CeO_2–Y_2O_3 nanocrystals, the intensity increase of the luminescence band caused by 5d → 4f transitions of Ce^{3+} ions is also accompanied by an increase in the intensity of the band caused by F^0-centers. However, the intensity of F^0 centers increases more slowly with increasing zirconium concentration than the intensity of Ce^{3+} luminescence, which suggests that during formation of oxygen vacancies in CeO_2-ZrO_2 nanocrystals, the process of electron capture by Ce^{4+} ions prevails over the capture of electrons by oxygen vacancies (Fig. 2.21).

These results are consistent with those previously obtained using other methods. Thus, in [51], it was shown by the methods of X-ray absorption fine structure

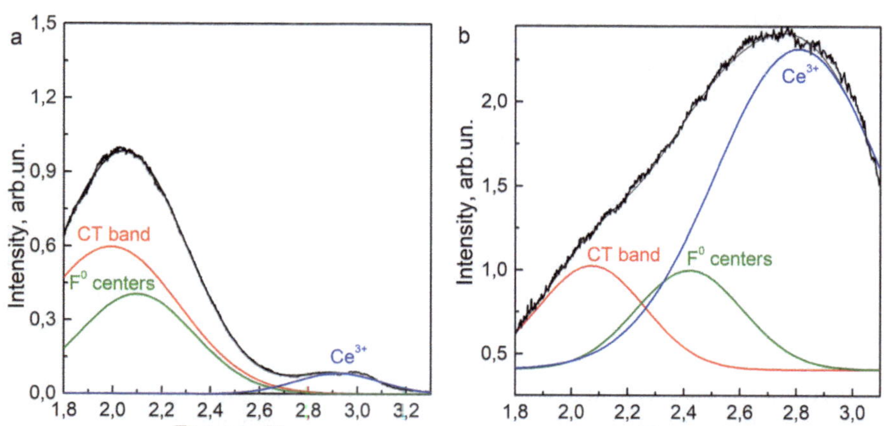

Fig. 2.20 Luminescence spectra of CeO_2-ZrO_2 nanocrystals ($T = 10$ K) with different zirconium content: (**a**) 10 at.%; (**b**) 20 at.%, after treatment in reducing environment

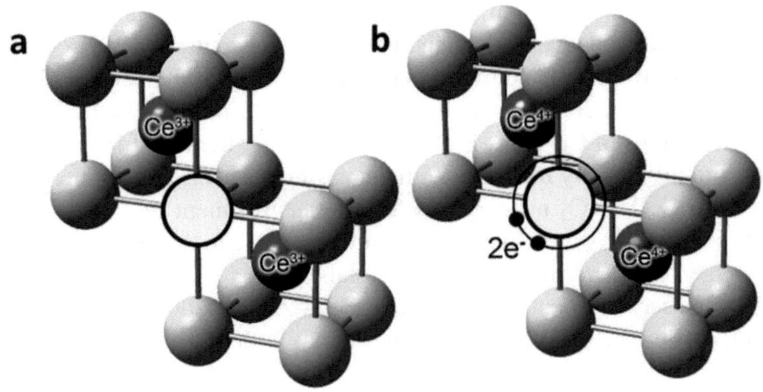

Fig. 2.21 The main types of defect centers in CeO_2-ZrO_2 nanocrystals: (**a**) Ce^{3+}-Vo-Ce^{3+} complexes; (**b**) F^0-centers

spectra (EXAFS) that the replacement of Ce^{4+} ions with Zr^{4+} ions leads to a significant increase in the number of oxygen vacancies in cerium oxide. According to [51], for the mixed $Ce_{0.5}Zr_{0.5}O_2$ oxide, a significant distortion of the oxygen sublattice around zirconium ions leads to an increase in the Zr-O distances for two of eight oxygen ions surrounding the Zr^{4+} ion (0.26 nm compared to 0.2312 nm for Ce-O in pure CeO_2). Longer Zr-O distances for some oxygen ions and lower bond energy reduce the energy of oxygen vacancy formation. This fact is confirmed by lower activation energy of oxygen diffusion in these materials, and therefore by the improvement of the catalytic properties of materials based on mixed oxides [52].

2.3.3 *Dependence of the Content of Ce^{3+}-Vo-Ce^{3+} and Re^{3+}-Vo-Ce^{3+} Complexes in Cerium Oxide Nanocrystals on the Type and Content of Impurity Ions*

In Sect. 2.1, it was shown that the transition from oxidizing to reducing treatment atmosphere, as well as a decrease in the size of nanocrystals from 50 nm to 2 nm, leads to an increase in the intensity of Ce^{3+} luminescence due to an increase of the content of oxygen vacancies and, therefore, Ce^{3+}-Vo-Ce^{3+} complexes in cerium oxide nanocrystals. As well as varying the treatment atmosphere or reducing the size of nanocrystals, presence of impurity ions (both isovalent and non-isovalent) leads to an increase in the intensity of Ce^{3+} luminescence band. As was shown in Sects. 2.3.1 and 2.3.2, this is due to the dominant role of Ce^{3+}-Vo-Re^{3+} complexes (where Re = Y, Tb, Eu) at high concentrations of aliovalent impurity ions (Y^{3+}, Tb^{3+}, Eu^{3+}, respectively), and decrease in the energy of the formation of oxygen vacancies, and therefore, an increase in the content of Ce^{3+}-Vo-Ce^{3+} complexes in the presence of isovalent impurity ions (in particular, Zr^{4+} ions).

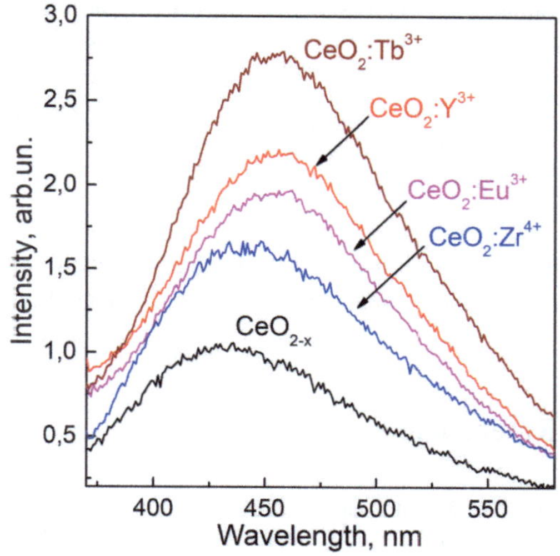

Fig. 2.22 Luminescence spectra of CeO_2 nanocrystals, CeO_2–Re_2O_3 (Re = Y, Eu, Tb), CeO_2–ZrO_2 (2 nm, C = 10 at.%)

A decrease in the size of mixed oxide nanocrystals should lead to an even higher increase in the content of Ce^{3+} ions due to the formation of additional Ce^{3+}-Vo-Ce^{3+} complexes, the content of which for small mixed oxide nanocrystals will be comparable to the content of Ce^{3+}-Vo-Re^{3+} complexes. However, the formation of these two types of complexes are not independent processes, since the impurity aliovalent ion can replace both Ce^{4+} ion and Ce^{3+} ion, so the total content of Ce^{3+} ions will depend, in particular, on the type of impurity ions.

Figure 2.22 shows the luminescence spectra of CeO_2, CeO_2–Re_2O_3 (Re = Y, Eu, Tb) and CeO_2–ZrO_2 (2 nm) nanocrystals. The content of impurity ions for all nanocrystals was 10 at.%. It can be seen from the spectra that the presence of impurity ions leads to an increase in the intensity of Ce^{3+} luminescence, and the highest intensity of Ce^{3+} luminescence and, therefore, the highest content of Ce^{3+} ions is observed for CeO_2-Tb_2O_3 nanocrystals (more than 2.5 times higher than for the CeO_{2-x} nanocrystals of similar size).

As in 50 nm nanocrystals with high concentrations of aliovalent Re^{3+} impurity ions, the main type of defects are Ce^{3+}-Vo-Re^{3+} complexes (their number is much higher than the number of Ce^{3+}-Vo-Ce^{3+} complexes), the content of Ce^{3+} ions in such nanocrystals at high concentrations of impurity ions (Re^{3+}) approximately corresponds to the content of impurity ions. This made it possible to compare the content of Ce^{3+} ions in all obtained nanocrystals (Table 2.2). As can be seen from the table, the content of Ce^{3+} ions in mixed oxide nanocrystals can reach up to 30–35 at.% while preserving the fluorite-like structure of the cerium oxide nanocrystal. It should be noted that the obtained values of the content of Ce^{3+} ions in nanocrystals are somewhat lower in comparison with the data obtained by the methods of X-ray photoelectron spectroscopy (so, according to our data, the content of Ce^{3+} ions in nanocrystals with a size of 2 nm is about 15%, and not 40–50%, as was shown in [21, 29]). However, this may be due to the partial reduction of $Ce^{4+} \rightarrow Ce^{3+}$ ions

Table 2.2 The content of Ce^{3+} ions in CeO_{2-x} nanocrystals depending on the size of the nanocrystal and the type of impurity ions (at impurity concentration of 10 at.%), determined by the intensity of the Ce^{3+} luminescence band

	CeO_{2-x}			CeO_2-ZrO_2,	CeO_2-Eu_2O_3,	CeO_2-Y_2O_3,	CeO_2-Tb_2O_3,
	50 nm (H_2)	10 nm	2 nm	2 nm	2 nm	2 nm	2 nm
Ce^{3+}, %	~1.5	~5	~15	~22	~25	~30	~35

under an action of X-ray irradiation when determining the content of Ce^{3+} ions in cerium oxide nanocrystals by the XPS method.

References

1. Scirè S, Palmisano L (2019) Cerium oxide (CeO_2): synthesis, properties and applications. Elsevier. https://doi.org/10.1016/C2017-0-02724-6
2. Orel ZC, Orel B (1994) Optical properties of pure CeO_2 and mixed CeO_2/SnO_2 thin film coatings. Phys Status Solidi B 186:K33–K36. https://doi.org/10.1002/pssb.2221860135
3. Marabelli F, Wachter P (1987) Covalent insulator CeO_2: optical reflectivity measurements. Phys Rev B 36:1238–1243. https://doi.org/10.1103/PhysRevB.36.1238
4. Goubin F, Rocquefelte X, Whangbo M-H (2004) Experimental and theoretical characterization of the optical properties of CeO_2, $SrCeO_3$, and Sr_2CeO_4 containing Ce^{4+} (f^0) ions. Chem Mater 16:662–669. https://doi.org/10.1002/chin.200417017
5. Fabris S, de Gironcoli S, Baroni S (2005) Taming multiple valency with density functionals: a case study of defective ceria. Phys Rev B 71:041102-1–041102-4. https://doi.org/10.1103/PhysRevB.71.041102
6. Nolan M, Fearon JE, Watson GW (2006) Oxygen vacancy formation and migration in ceria. Solid State Ionics 177:3069–3074. https://doi.org/10.1016/j.ssi.2006.07.045
7. Loschen C, Carrasco J, Neyman KM, Illas F (2007) First-principles LDA+U and GGA+U study of cerium oxides: dependence on the effective U parameter. Phys Rev B 75:035115. https://doi.org/10.1103/PhysRevB.75.035115
8. Da Silva JL, Ganduglia-Pirovano MV, Sauer J, Bayer V, Kresse G (2007) Hybrid functionals applied to rare-earth oxides: the example of ceria. Phys Rev B 75:045121. https://doi.org/10.1103/PhysRevB.75.045121
9. Skorodumova NV, Simak SI, Lundqvist BI, Abrikosov IA, Johansson B (2002) Quantum origin of the oxygen storage capability of ceria. Phys Rev Lett 89:166601. https://doi.org/10.1103/PhysRevLett.89.166601
10. Liu LZ, Allen JW, Gunnarsson O, Christensen NE, Andersen OK (1992) A-Γ transition in Ce: a detailed analysis of electron spectroscopy. Phys Rev B 45:8934. https://doi.org/10.1103/PhysRevB.45.8934
11. Johansson B (1974) The α-γ transition in cerium is a Mott transition. Philos Mag 30:469–482. https://doi.org/10.1080/14786439808206574
12. Johansson B, Abrikosov IA, Aldén M, Ruban AV, Skriver HL (1995) Calculated phase diagram for the $\gamma \rightleftharpoons \alpha$ transition in Ce. Phys Rev Lett 74:2335. https://doi.org/10.1103/PhysRevLett.74.2335
13. Aškrabić S, Dohčević-Mitrović ZD, Araújo VD, Ionita G, de Lima MM Jr, Cantarero A (2013) F-centre luminescence in nanocrystalline CeO_2. J Phys D Appl Phys 46:495306. https://doi.org/10.1088/0022-3727/46/49/495306
14. Tiseanu C, Parvulescu VI, Sanchez-Dominguez M, Boutonnet M (2012) Temperature induced conversion from surface to “bulk” sites in Eu^{3+}-impregnated CeO_2 nanocrystals. J Appl Phys 112:013521

15. Wei X, Pan W, Cheng L, Li B (2009) Atomistic calculation of association energy in doped ceria. Solid State Ionics 180:13–17. https://doi.org/10.1016/j.ssi.2008.10.019
16. Yoshida H, Deguchi H, Miura K, Horiuchi M, Inagaki T (2001) Investigation of the relationship between the ionic conductivity and the local structures of singly and doubly doped ceria compounds using EXAFS measurement. Solid State Ionics 140:191–199. https://doi.org/10.1016/S0167-2738(01)00854-2
17. Nakayama M, Martin M (2009) First-principles study on defect chemistry and migration of oxide ions in ceria doped with rare-earth cations. Phys Chem Chem Phys 11:3241. https://doi.org/10.1039/B900162J
18. Yen WM, Shionoya S, Yamamoto H (1997) Phosphor handbook (Laser & Optical Science & Technology). CRC Press, London
19. Esch F, Fabris S, Zhou L, Montini T, Africh C, Fornasiero P, Rosei R (2005) Electron localization determines defect formation on ceria substrates. Science 309:752–755. https://doi.org/10.1126/science.1111568
20. Dutta P, Pal S, Seehra MS, Shi Y, Eyring EM, Ernst RD (2006) Concentration of Ce^{3+} and oxygen vacancies in cerium oxide nanoparticles. Chem Mater 18:5144–5146. https://doi.org/10.1021/cm061580n
21. Deshpande S, Patil S, Kuchibhatla SV, Seal S (2005) Size dependency variation in lattice parameter and valency states in nanocrystalline cerium oxide. Appl Phys Lett 87:133113. https://doi.org/10.1063/1.2061873
22. Shcherbakov AB, Zholobak NM, Ivanov VK (2020) Biological, biomedical and pharmaceutical applications of cerium oxide. In: cerium oxide (CeO_2): synthesis, properties and applications. Elsevier, pp 279–358. https://doi.org/10.1016/B978-0-12-815661-2.00008-6
23. Cafun JD, Kvashnina KO, Casals E, Puntes VF, Glatzel P (2013) Absence of Ce^{3+} sites in chemically active colloidal ceria nanoparticles. ACS Nano 7:10726–10732. https://doi.org/10.1021/nn403542p
24. Figgis BN, Hitchman MA (2000) Ligand field theory and its applications. Wiley-VCH, New York
25. Aull BF, Jenssen HP (1986) Impact of ion-host interactions on the 5d-to-4f spectra of lanthanide rare-earth-metal ions. I. A phenomenological crystal-field model. Phys Rev B 34:6640. https://doi.org/10.1103/PhysRevB.34.6640
26. Nakazawa E (2002) The lowest 4f-to-5d and charge-transfer transitions of rare earth ions in YPO_4 hosts. J Luminescence 100:89–96. https://doi.org/10.1016/S0022-2313(02)00429-5
27. Spanier JE, Robinson RD, Zhang F, Chan SW, Herman IP (2001) Size-dependent properties of CeO_{2-y} nanoparticles as studied by Raman scattering. Phys Rev B 64:245407. https://doi.org/10.1103/PhysRevB.64.245407
28. Tsunekawa S, Sivamohan R, Ito S, Kasuya A, Fukuda T (1999) Structural study on monosize CeO_{2-x} nanoparticles. Nanostruct Mater 11:141–147. https://doi.org/10.1016/S0965-9773(99)00027-6
29. Hao X, Yoko A, Chen C, Inoue K, Saito M, Seong G, Ikuhara Y (2018) Atomic-scale valence state distribution inside ultrafine CeO_2 Nanocubes and its size dependence. Small 14:1802915. https://doi.org/10.1002/smll.201802915
30. Tsunekawa S, Ishikawa K, Li ZQ, Kawazoe Y, Kasuya A (2000) Origin of anomalous lattice expansion in oxide nanoparticles. Phys Rev Lett 85:3440. https://doi.org/10.1103/PhysRevLett.85.3440
31. Li G, Boerio-Goates J, Woodfield BF, Li L (2004) Evidence of linear lattice expansion and covalency enhancement in rutile TiO_2 nanocrystals. Appl Phys Lett 85:2059–2061. https://doi.org/10.1063/1.1790596
32. Hailstone RK, DiFrancesco AG, Leong JG, Allston TD, Reed KJ (2009) A study of lattice expansion in CeO_2 nanoparticles by transmission electron microscopy. J Phys Chem C 113:15155–15159. https://doi.org/10.1021/jp903468m
33. Li L, Yang HK, Moon BK, Fu Z, Guo C, Jeong JH, Lee HS (2009) Photoluminescence properties of CeO_2: Eu^{3+} nanoparticles synthesized by a sol-gel method. J Phys Chem C 113:610–617. https://doi.org/10.1021/jp808688w

34. Viagin O, Masalov A, Ganina I, Malyukin Y (2009) Mechanism of energy transfer in $Sr_2CeO_4:Eu^{3+}$ phosphor. Opt Mater 31:1808–1810. https://doi.org/10.1016/j.optmat.2008.12.038
35. Alpaslan E, Yazici H, Golshan NH, Ziemer KS, Webster TJ (2015) pH-dependent activity of dextran-coated cerium oxide nanoparticles on prohibiting osteosarcoma cell proliferation. ACS Biomater Sci Eng 1:1096–1103. https://doi.org/10.1021/acsbiomaterials.5b00194
36. Jørgensen CK (2007) Electron transfer spectra, pp 101–158
37. Chai C, Yang S, Liu Z, Liao M, Chen N (2003) Violet/blue photoluminescence from CeO_2 thin film. Chin Sci Bull 48:1198–1200. https://doi.org/10.1007/BF03183935
38. Mochizuki S, Fujishiro F (2009) The photoluminescence properties and reversible photoinduced spectral change of CeO_2 bulk, film and nanocrystals. Phys Status Solidi (b) 246:2320–2328. https://doi.org/10.1002/pssb.200844419
39. Berberan-Santos MN, Bodunov EN, Valeur B (2005) Mathematical functions for the analysis of luminescence decays with underlying distributions: 2. Becquerel (compressed hyperbola) and related decay functions. Chem Phys 317:57–62. https://doi.org/10.1016/j.chemphys.2005.05.026
40. Adirovitch EI (1956) La formule de Becquerel et la loi élémentaire du déclin de la luminescence des phosphores cristallins. J Phys Radium 17:705–707
41. Radović M, Stojadinović B, Tomić N (2014) Investigation of surface defect states in $CeO_{2-\gamma}$ nanocrystals by scanning tunneling microscopy/spectroscopy and ellipsometry. J Appl Phys 116:234305. https://doi.org/10.1063/1.4904516
42. Henderson MA, Perkins CL, Engelhard MH, Thevuthasan S, Peden CHF (2003) Redox properties of water on the oxidized and reduced surfaces of CeO_2 (111). Surf Sci 526:1–18. https://doi.org/10.1016/S0039-6028(02)02657-2
43. Avram D, Gheorghe C, Rotaru C, Cojocaru B, Florea M, Parvulescu V, Tiseanu C (2014) Lanthanide–lanthanide and lanthanide–defect interactions in co-doped ceria revealed by luminescence spectroscopy. J Alloys Compd 616:535–541. https://doi.org/10.1016/j.jallcom.2014.07.059
44. van Herle J, Horita T, Kawada T, Sakai N, Yokokawa H, Dokiya M (1996) Sintering behaviour and ionic conductivity of yttria-doped ceria. J Eur Ceram Soc 16:961–973. https://doi.org/10.1016/0955-2219(96)00012-X
45. Andersson DA, Simak SI, Skorodumova NV, Abrikosov IA, Johansson B (2006) Optimization of ionic conductivity in doped ceria. Proc Natl Acad Sci 103:3518–3521. https://doi.org/10.1073/pnas.0509537103
46. Balducci G, Kašpar J, Fornasiero P, Graziani M, Islam MS (1998) Surface and reduction energetics of the CeO_2–ZrO_2 catalysts. J Phys Chem B 102:557–561. https://doi.org/10.1021/jp972400n
47. Fornasiero P, Balducci G, Meriani S, di Monte R, Graziani M (1996) Metal-loaded CeO_2–ZrO_2 solid solutions as innovative catalysts for automotive catalytic converters. Catal Today 29:47–52. https://doi.org/10.1016/0920-5861(95)00261-8
48. Bekyarova E, Fornasiero P, Kašpar J, Graziani M (1998) CO oxidation on Pd/CeO_2–ZrO_2 catalysts. Catal Today 45:179–183. https://doi.org/10.1016/S0920-5861(98)00212-0
49. Vlaic G, di Monte R, Fornasiero P, Fonda E, Kašpar J, Graziani M (1998) The CeO_2–ZrO_2 system: redox properties and structural relationships. Catal Automotive Pollut Control IV 116:185–195
50. Trubelja MF, Stubican VS (1991) Ionic conductivity in the hafnia–R_2O_3 systems. Solid State Ionics 49:89–97. https://doi.org/10.1016/0167-2738(91)90073-K
51. Vlaic G, Fornasiero P, Geremia S, Kašpar J, Graziani M (1997) Relationship between the zirconia-promoted reduction in the Rh-loaded $Ce_{0.5}Zr_{0.5}O_2$ mixed oxide and the Zr–O local structure. J Catal 168:386–392. https://doi.org/10.1006/jcat.1997.1644
52. Dong F, Suda A, Tanabe T, Nagai Y, Sobukawa H, Shinjoh H, Duprez D (2004) Dynamic oxygen mobility and a new insight into the role of Zr atoms in three-way catalysts of Pt/CeO_2–ZrO_2. Catal Today 93:827–832. https://doi.org/10.1016/j.cattod.2004.06.076

Chapter 3
Antioxidant Properties of CeO_{2-x} Nanocrystals

Abstract This chapter explores the antioxidant and prooxidant properties of cerium oxide nanocrystals (CeO_{2-x}) and their interaction with reactive oxygen species (ROS). Since their discovery as self-regenerating antioxidants, CeO_{2-x} nanocrystals have been shown to mimic the activity of natural enzymes, including superoxide dismutase and catalase, thereby regulating intracellular redox balance. The chapter reviews mechanisms proposed for their unusual activity, including roles of oxygen vacancies, Ce^{3+}/Ce^{4+} redox cycling, partial dissolution, and Ce^{3+}-V_0-Ce^{3+} complexes. Using luminescence spectroscopy, it is demonstrated that the 5d → 4f luminescence of Ce^{3+} ions serves as a sensitive probe of redox dynamics, enabling real-time monitoring of ROS decomposition and subsequent regeneration of Ce^{3+} content in nanocrystals. The catalytic activity of nanoceria follows Michaelis–Menten kinetics, consistent with the presence of discrete catalytic sites, most likely associated with Ce^{3+}-V_0-Ce^{3+} complexes. Distinct behaviors are revealed for different ROS (H_2O_2, ClO^-, $O_2^{\bullet -}$, and $^{\bullet}OH$), for which dopant-induced defect engineering can modulate antioxidant or prooxidant effects. Notably, high oxidant concentrations can induce long-term oscillations in the Ce^{3+}/Ce^{4+} ratio, reminiscent of non-equilibrium chemical oscillations. The results highlight both the biomedical potential and mechanistic complexity of CeO_{2-x} nanocrystals as nanozymes, providing insights into their application in oxidative stress-related diseases and redox-based nanomedicine.

The unique biological properties of cerium oxide nanocrystals were first discovered in 2005, when two papers were published by different scientific groups, the first of which [1] has shown that cerium oxide nanocrystals can extend the lifespan of neurons in cell culture by more than five times from 20 days to 100 days or more; and the second [2] has shown that cerium oxide nanocrystals in a pool of cancerous and healthy cells selectively accumulated in healthy cells and protected them from the effects of X-ray radiation providing thereby the possibility of using these nanocrystals in radiotherapy of tumors. These papers marked the beginning of biological research on nanocrystalline cerium oxide as an antioxidant [3–10].

V. Seminko, P. Maksimchuk, *Cerium Oxide Nanocrystals (CeO_{2-x})*,
https://doi.org/10.1007/978-3-032-14130-9_3

As known, living cells contain large amounts of reactive oxygen species (ROS), such as superoxide anions $O_2^{\bullet-}$, hydrogen peroxide H_2O_2, and hydroxyl radicals $^{\bullet}OH$, which are formed during cellular respiration and play an important role in transmission of intracellular signals and immune response. However, if ROS concentration exceeds a certain limit, the oxidation of lipid membranes and other intracellular components occurs, which, according to one of the theories of aging, can eventually lead to aging of an organism as a whole [11–15].

ROS can change from one form to another according to the following scheme:

$$O_2 \xrightarrow{e^-} O_2^- \xrightarrow{e^-} H_2O_2 \xrightarrow{e^-} {}^{\cdot}OH + OH^- \xrightarrow{e^-} 2H_2O$$

Superoxide anion $O_2^{\bullet-}$ shows weak redox activity and also has no ability to pass through lipid membranes. Hydrogen peroxide H_2O_2 is an intracellular signaling molecule [16, 17], and it has the ability to pass through biological membranes and plays an important role in the formation of other ROS, which have stronger redox activity, first of all, hydroxyl radicals ($^{\bullet}OH$). Hydroxyl radicals exhibit significant redox activity, as a result of which they are able to cause more damage to biological systems than any other ROS [18, 19]. Hydroxyl radicals are formed from hydrogen peroxide in a reaction catalyzed by metal ions (Fe^{2+} or Cu^+) bound in complexes with various proteins or other molecules (Fenton's reaction) [20]:

$$H_2O_2 + Cu^+ / Fe^{2+} \rightarrow {}^{\cdot}OH + OH^- + Cu^{2+} / Fe^{3+} \tag{3.1}$$

Ions of transition metals [21–23] which can be released from proteins can interact with superoxide anions ($O_2^{\bullet-}$) providing formation of hydroxyl radicals as well.

To control the level of ROS, a living cell uses enzymes such as superoxide dismutase (which converts superoxide into hydrogen peroxide) and catalase (which converts hydrogen peroxide into water and oxygen), as well as antioxidant molecules such as ascorbic acid, which unlike enzymes are not capable of reversible action (their recovery after interaction with ROS requires the participation of glutathione or other biological molecules) [24–26].

It was shown that cerium oxide, like enzymes in a living cell, is a unique self-regenerating antioxidant capable of simultaneously converting superoxide into hydrogen peroxide and hydrogen peroxide into oxygen and water, i.e., acting like superoxide dismutase [27] and catalase [28]. Later, similar activity was also found for other nanocrystals with ions of variable valence. These studies actually gave birth to a new interdisciplinary field called nanoenzymology [29, 30]. Nanoparticles with enzyme-like activity (nanozymes) attract sufficient interest from scientific communities worldwide. Of particular interest are nanoparticles that, depending on the external conditions and parameters of the nanoparticle (size, surface morphology, etc.), can demonstrate an effect similar to the effect of various enzymes [31–33].

Oxide nanocrystals (and, in particular, cerium oxide nanocrystals CeO_{2-x}) were among the first for which an activity similar to the activity of enzymes in a living cell was found, and they still remain one of the most promising materials from the point of view of the variety of possibilities for realizing their anti-/prooxidant activity, as well as due to the number of biomedical studies of these nanoparticles. Recent studies have shown that CeO_{2-x} nanocrystals can demonstrate both catalase- and superoxide dismutase- (SOD-), as well as oxidase- [34] and peroxidase-like [35] activity, the first two provide antioxidant (neutralization of ROS), and the last two prooxidant (oxidation of biological molecules) action of nanocrystals. In addition, CeO_{2-x} nanocrystals are characterized by the presence of photocatalytic prooxidant activity, which is observed when the nanocrystals are irradiated with radiation with a wavelength close to the width of the optical band gap of the nanocrystal (~3 eV) [36].

The study of ROS interaction processes with CeO_{2-x} nanoparticles is complicated by the influence of a large number of parameters (pH, ROS concentration, Ce^{3+}/Ce^{4+} ratio) on the outcome and dynamics of such interaction. Until now, most authors used X-ray photoelectron spectroscopy (XPS) methods to determine the change in the Ce^{3+}/Ce^{4+} ratio during the interaction of nanoparticles with ROS. In [27], two types of cerium oxide nanocrystals were studied: one with a low content and the other with a high content of Ce^{3+} ions on the surface. It was shown that the method of synthesis of nanoparticles and the chemical properties of the surface can significantly change the SOD-like activity of nanocrystals. It was also shown that for nanoparticles with high concentration of Ce^{3+} ions, the chemical properties of the surface changed during oxidation of Ce^{3+} ions to Ce^{4+} with hydrogen peroxide. This change in chemical properties of the surface correlated with a decrease in SOD-like activity.

Interestingly, according to the results of many authors, catalase-like activity of cerium oxide nanocrystals is inversely related to SOD-like activity. In [28], it was found that the catalase-like activity decreases with an increase in the concentration of Ce^{3+} ions on the surface of the nanoparticle, while SOD-like activity shows opposite behavior. The catalase-like activity of cerium oxide nanocrystals is much weaker than that of the catalase enzyme, and therefore it was assumed that the catalytic activity of cerium oxide nanocrystals can be low in vivo [28]. However, since peroxides are the most abundant and persistent ROS in vivo, a catalyst reducing peroxide level can be used during inflammation or to prevent reactions catalyzed by Fenton reactions.

In [37], the effect of phosphate ions on the activity of cerium oxide was shown, in particular, phosphate can change SOD-like activity to catalase-like in vitro changing the dominant valence state of cerium ions.

In general, the extremely high antioxidant activity of cerium oxide nanocrystals in biological conditions has been repeatedly demonstrated by various research groups. In [38], the SOD-like activity of cerium oxide nanocrystals was investigated in living cells, in particular in brain cells, and it was demonstrated that the cerium oxide nanocrystals decrease the level of superoxide in cells ex vivo. The authors of [39] show that cerium oxide nanocrystals with a size of 3.8 nm covered with a thin layer of oleic acid have an antioxidant activity nine times higher than commercial antioxidant Trolox™. A number of studies were devoted to the possibility of using cerium oxide nanocrystals to treat diseases such as cancer, Alzheimer's disease, and cardiovascular diseases. In [4], it was shown that the use of nanocrystalline cerium oxide can significantly increase the effectiveness of anticancer radiotherapy, since CeO_{2-x} nanoparticles are selectively accumulated in normal cells (without entering cancer cells) protecting them from an increase in the concentration of free radicals during tumor irradiation. The authors [3] showed that cerium oxide provides effective protection of rat spinal cord neurons against oxidative stress. A reduction in the degree of ischemic brain damage by ROS in the presence of cerium oxide nanocrystals was shown in [40].

Today, there are a number of models developed to explain the unique antioxidant activity of cerium oxide nanocrystals. They can be divided into four categories (Fig. 3.1). The first of them relies on the antioxidant properties of cerium oxide nanocrystals to high oxygen capacity of this material [41]. According to this model, oxygen vacancies play the main role in the antioxidant activity of cerium oxide nanocrystals. According to the second model, the antioxidant properties of CeO_{2-x} nanoparticles are caused by $Ce^{3+} \leftrightarrow Ce^{4+}$ redox processes on the nanocrystal surface [27, 28]. The third model states that there are no Ce^{3+} ions in nanocrystals at all [42, 44–46] (for instance, the authors of [47] have not detected Ce^{3+} ions in cerium oxide nanocrystals), while antioxidant activity is provided by ions released from nanoparticles when they are partially dissolved in water or other biological fluids [48]. According to the fourth model, the antioxidant activity of cerium oxide nanocrystals is provided not by individual Ce^{3+}/Ce^{4+} ions or oxygen vacancies, but rather by Ce^{3+}-Vo-Ce^{3+} complexes (where Vo is an oxygen vacancy) [43].

Each of these theories is not without certain drawbacks, although in general most researchers today support the second model, according to which the interaction of ROS with cerium oxide nanocrystals occurs through oxidation/reduction of Ce^{3+}/Ce^{4+} ions on the surface of the nanocrystal. To confirm that Ce^{3+}/Ce^{4+} ions, and not

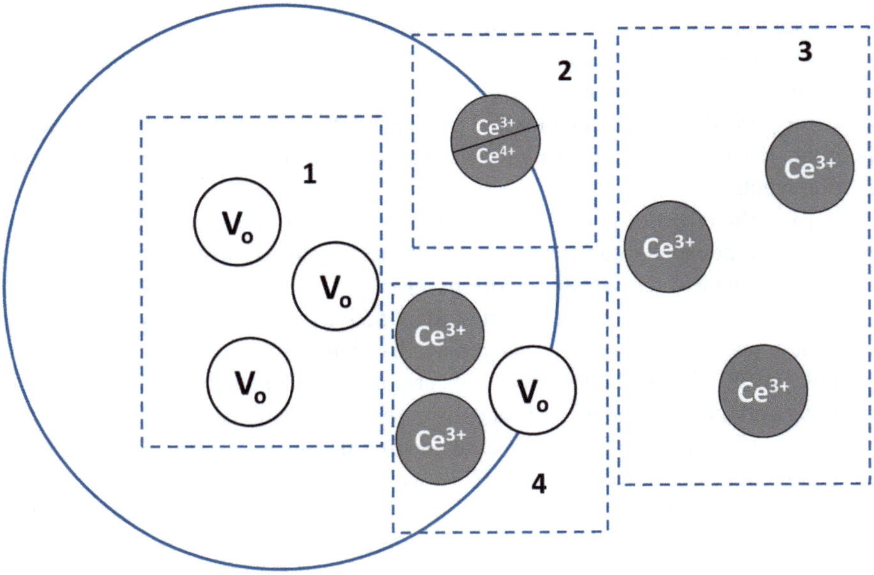

Fig. 3.1 Types of defects determining the antioxidant activity of cerium oxide nanocrystals according to different authors: (1) oxygen vacancies [41]; (2) Ce^{3+}/Ce^{4+} ions [27, 28]; (3) Ce^{3+} ions released from nanoparticle during its partial dissolution [42]; (4) Ce^{3+}-Vo-Ce^{3+} complexes [43]

oxygen vacancies, play the main role, the authors of [49] obtained mixed CeO_2-Sm_2O_3 nanocrystals. The content of oxygen vacancies in them was higher than in cerium oxide nanocrystals, while the concentration of Ce^{3+} ions was lower, since part of the Ce^{3+} ions was replaced by samarium ions. For such nanocrystals, the antioxidant activity was lower than for cerium oxide that was interpreted by the authors as an argument in favor of their model. However, another scientific group has shown that the presence of zirconium ions in cerium oxide nanocrystals, on the contrary, leads to a significant increase in antioxidant activity, which calls into question the exclusive role of Ce^{3+}/Ce^{4+} ions and emphasizes the role of the content of oxygen vacancies in the antioxidant effect of CeO_{2-x} nanocrystals [41].

In addition, the interaction of Ce^{3+} ions, in particular, with hydrogen peroxide, should not lead to catalase-like activity, but to generation of hydroxyl radicals according to the Fenton mechanism, which are much more reactive and dangerous for living cells [20]. In order to bypass the possibility of the Fenton reaction, two options were proposed: according to the first, the decomposition of hydrogen peroxide is not accompanied by $Ce^{3+} \rightarrow Ce^{4+}$ transition, but on the contrary, $Ce^{4+} \rightarrow Ce^{3+}$ one, i.e., hydrogen peroxide should act not as an oxidant, but as a reducing agent [50]:

$$H_2O_2 + 2Ce^{4+} + 2OH^- \rightarrow 2H_2O + O_2 + 2Ce^{3+} \tag{3.2}$$

According to the second, the generation of hydroxyl radicals is indeed observed, but later these radicals react with hydrogen peroxide to form superoxide anions [39]:

$$H_2O_2 + Ce^{3+} \rightarrow Ce^{4+} + OH^- + {}^{\cdot}OH \tag{3.3}$$

$${}^{\cdot}OH + H_2O_2 \rightarrow H_2O + H^+ + O_2^{\,-} \tag{3.4}$$

In general, despite a significant number of publications on this topic, there is no consensus on the mechanisms of antioxidant activity of cerium oxide nanocrystals [51–53].

3.1 Luminescence Spectroscopy Methods for Studying the Dynamics of ROS Interaction with CeO_{2-x} Nanocrystals

To date, despite an extremely large number of works devoted to the study of the antioxidant properties of cerium oxide nanocrystals, there is no clear understanding of the mechanisms that lead to the abnormally strong antioxidant activity of these nanocrystals, and, in particular, to their ability to regenerate their antioxidant properties when interacting with ROS in the biological environment. Most results in this field were obtained for rather complex biological systems making it even more difficult to understand the mechanisms of regeneration of the Ce^{3+}/Ce^{4+} ratio in nanocrystals during interaction with ROS.

A significant problem that prevents the use of these materials in medical practice is the impossibility of controlling the processes of interaction of cerium oxide nanocrystals with ROS in a living cell. The presence of 5d → 4f luminescence of Ce^{3+} ions, as will be shown later, makes it possible to achieve significant progress in solving the problem of controlling the processes of nanoparticle interaction with ROS, and revealing the physical mechanisms of the antioxidant action of CeO_{2-x} nanoparticles. In fact, the 5d → 4f luminescence of Ce^{3+} ions opens the possibility of monitoring the dynamics of the interaction between nanoparticles and oxidant molecules using optical spectroscopy methods.

Today, the main methods used to study the change in the valence state of cerium ions in the processes of interaction of CeO_{2-x} nanocrystals with ROS are the methods of X-ray photoelectron spectroscopy and X-ray absorption spectroscopy [27, 28, 37]. These methods require X-ray irradiation of the samples in vacuum chambers, which can lead to experimentally obtained values of the content of Ce^{3+} and Ce^{4+} ions, which differ significantly from the real values for nanoparticles in colloidal solution. In turn, the methods of luminescence spectroscopy do not require high vacuum conditions and high-energy irradiation, and, moreover, provide an opportunity to study the processes of the interaction of cerium oxide with ROS in dynamics. The presence of 5d → 4f luminescence of Ce^{3+} ions in cerium oxide nanocrystals opens the possibility of using luminescence spectroscopy methods to study the processes of interaction of CeO_{2-x} nanocrystals with different types of ROS (Sect. 3.2 is devoted to this problem) and also makes possible to reveal the nature of the mechanisms of regeneration of antioxidant properties of nanocrystals after interaction with ROS.

As was shown in Chap. 2, the luminescence spectra of cerium oxide nanocrystals (50 nm) at room temperature contain a band formed by 5d → 4f transitions of Ce^{3+} ions (Ce^{3+} luminescence band). The same band is observed in colloidal solutions of small cerium oxide nanoparticles (2 nm and 10 nm), moreover, a decrease in the size of nanoparticles leads to a significant increase in the intensity of their luminescence.

In subsequent experiments, it was shown that the addition of oxidants (in particular, ROS) to colloidal solutions of nanoparticles of cerium oxide leads to decrease in the intensity of Ce^{3+} luminescence band due to the oxidation of part of the Ce^{3+} ions ($Ce^{3+} \rightarrow Ce^{4+}$). In Fig. 3.2, the results for hydrogen peroxide are presented; however, the same effects were also observed for other ROS. The next section will highlight the differences in the processes of interaction of different types of ROS with cerium oxide nanoparticles, while this section focuses on the mechanisms common for different types of ROS.

As can be seen from Fig. 3.2, the addition of hydrogen peroxide leads to decrease in the intensity of Ce^{3+} luminescence with subsequent slow recovery. The luminescence spectra before the addition of hydrogen peroxide and 120 h after the addition of hydrogen peroxide are practically the same. Studies of changes in the concentration of hydrogen peroxide in aqueous solution (given in detail in Sect. 3.2) show that a decrease in the intensity of Ce^{3+} luminescence is accompanied by a gradual decrease in the content of hydrogen peroxide in aqueous solutions. When the

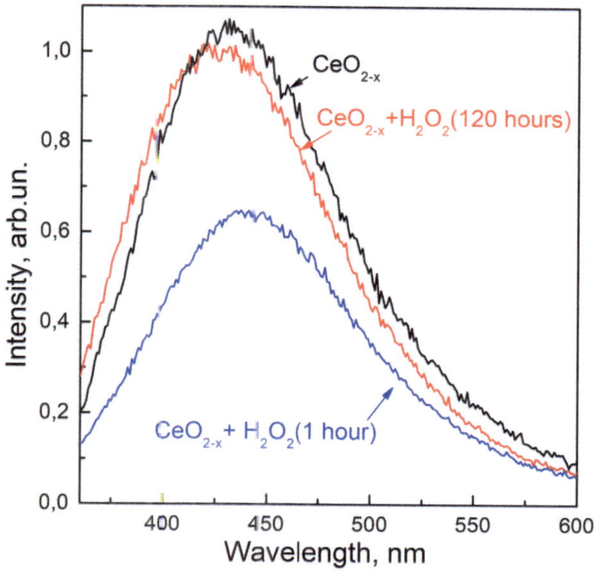

Fig. 3.2 Dynamics of Ce^{3+} luminescence intensity at interaction with ROS (luminescence spectra of colloidal solutions of CeO_{2-x} nanocrystals (2 nm) before and after adding hydrogen peroxide are shown). (Reprinted with permission from *Chemical Physics Letters*, vol. 767, V. Seminko, P. Maksimchuk, G. Grygorova, E. Okrushko, O. Sedyh, O. Avrunin, V. Semenets, Y. Malyukin, Switching the type of redox activity of colloidal nanoceria by Re^{3+} (Re = Y, Eu, Tb) doping, pp. 138363, 2021. Copyright 2021 Elsevier)

content of hydrogen peroxide in the aqueous solution becomes negligibly small, the process of recovery of Ce^{3+} luminescence intensity occurs until reaching the initial intensity observed before the addition of hydrogen peroxide. Since the intensity of Ce^{3+} luminescence is determined by the content of Ce^{3+} ions in nanocrystals, it can be stated that the process of decomposition of hydrogen peroxide by cerium oxide nanoparticles is accompanied by prevalence of $Ce^{3+} \rightarrow Ce^{4+}$ process, while after the end of the decomposition process, the initial content of Ce^{3+} ions in nanocrystals is recovered. This effect was previously observed in a number of biological experiments, which enabled the authors [54, 55] to talk about the ability of cerium oxide nanocrystals to regenerate their antioxidant properties when interacting with ROS. The fact that this process can be monitored by the change in the intensity of Ce^{3+} luminescence opens wide opportunities for studying the mechanisms of this effect.

The general dynamics of Ce^{3+} luminescence intensity, as already was noted above, is similar for different types of ROS (Fig. 3.3) and consists of an initial stage, which is accompanied by a decrease in the intensity of Ce^{3+} luminescence and corresponds to the process of decomposition of ROS by nanoparticles (for some ROS, in particular, hypochlorite anions (Fig. 3.3b), this stage turns into a stage in which the intensity of Ce^{3+} luminescence remains almost constant). After the end of ROS decomposition (in Fig. 3.3, it is approximately 50 h for hydrogen peroxide and

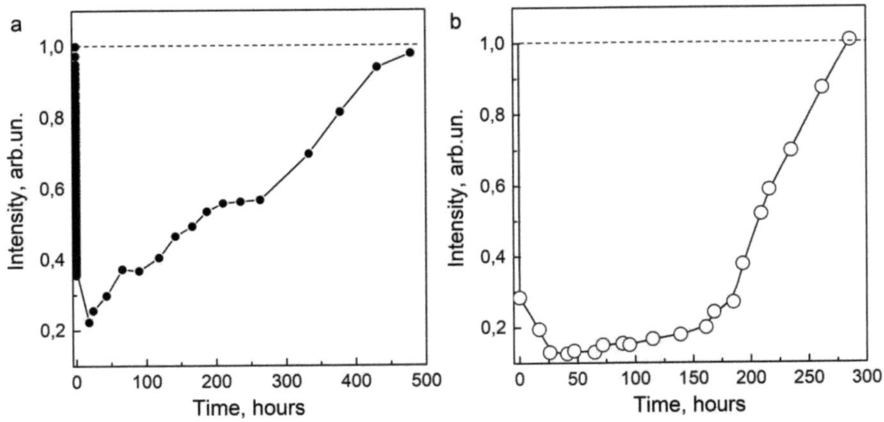

Fig. 3.3 General dynamics of changes in Ce^{3+} luminescence intensity of CeO_{2-x} nanocrystals (2 nm) after addition of hydrogen peroxide (**a**) and sodium hypochlorite (**b**). λ_{exc} = 325 nm, t = 22 °C. The initial concentration of ROS is 16 mM

170 h for hypochlorite anions), the stage of recovery of the initial intensity of Ce^{3+} luminescence begins. While the duration of the first stage depends on the initial concentration of ROS in the solution (the higher the concentration of ROS, the longer this stage is), the duration of the recovery stage of Ce^{3+} luminescence does not depend on the concentration of ROS and (as will be shown below) is determined solely by the parameters of the nanoparticle and the action of external factors (temperature, irradiation).

The key parameter that determines the duration of the recovery stage of the initial intensity of Ce^{3+} luminescence (and, therefore, the initial content of Ce^{3+} ions in the cerium oxide nanocrystal) is the temperature of the aqueous solution. Figure 3.4 shows the dynamics of recovery of Ce^{3+} luminescence intensity after interaction with hydrogen peroxide for CeO_{2-x} nanocrystals (2 nm) under conditions of temperature variation. It can be seen that the temperature increased only by 30°C (from 22 to 52 °C) leads to reduction in the recovery time of Ce^{3+} luminescence from 120 h to 5 h. The strong dependence of the Ce^{3+} luminescence recovery process on temperature suggests that this process requires thermal activation.

One of the key parameters of CeO_{2-x} nanocrystals is their ability to self-regenerate upon interaction with ROS, which, in turn, is due to the reversible switching of the valence of the cerium ion $Ce^{3+} \leftrightarrow Ce^{4+}$. These two effects can be explained if we assume that $Ce^{4+} \rightarrow Ce^{3+}$ reduction occurs synchronously with oxidation to provide the continuity of the antioxidant effect of the nanocrystal. One of the possible mechanisms of $Ce^{4+} \rightarrow Ce^{3+}$ recovery is water splitting at the oxidized Ce^{4+}-Vo-Ce^{4+} sites (Fig. 3.5).

The processes of catalytic decomposition of water by nanocrystals of cerium oxide were previously shown in a number of publications [56–58]. The fact that oxygen occupies the position of an oxygen vacancy on the surface of a nanocrystal should give a significant gain in energy. Due to the large oxygen non-stoichiometry

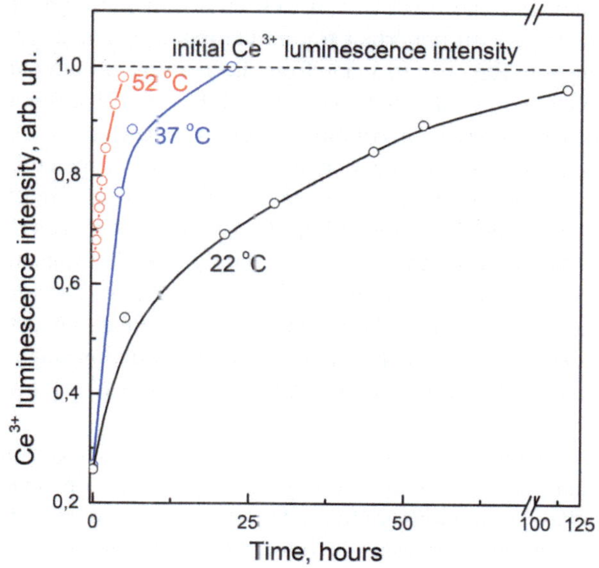

Fig. 3.4 Dynamics of recovery of the initial intensity of Ce^{3+} luminescence for CeO_{2-x} nanoparticles (2 nm) depending on the temperature

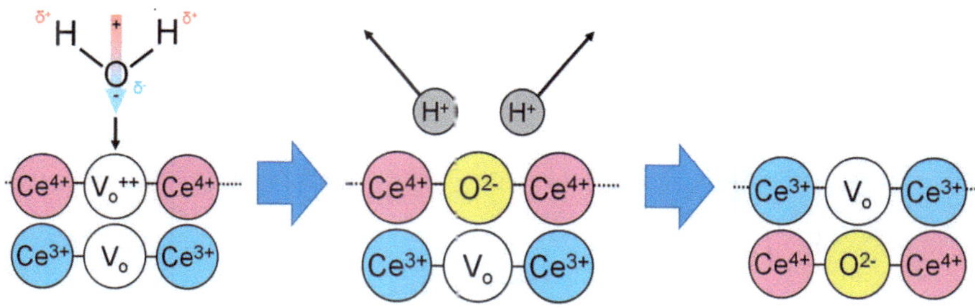

Fig. 3.5 A possible mechanism of regeneration of $Ce^{4+} \rightarrow Ce^{3+}$ on the surface of CeO_{2-x} nanoparticles involving water splitting on the surface Ce^{4+}-Vo-Ce^{4+} sites

of nanocrystals, oxygen obtained as a result of the catalytic decomposition on the surface of cerium oxide nanocrystals can diffuse into the nanoparticle over time. Thus, the reduction of $Ce^{4+} \rightarrow Ce^{3+}$ surface ions and the oxidation of $Ce^{3+} \rightarrow Ce^{4+}$ ions inside the CeO_{2-x} nanoparticle occur at the same time. Based on this, it is possible to associate the gradual decrease in the intensity of Ce^{3+} luminescence of nanocrystals upon addition of an oxidant with the gradual accumulation of oxygen by the nanoparticle and the prolonged stage of recovery after the complete decomposition of the oxidant with the gradual release of oxygen from the nanocrystal. In full accordance with this assumption, the last stage is accompanied by a gradual increase in the concentration of dissolved oxygen in the aqueous solution and is marked by the formation of bubbles on the walls of the cuvettes.

The conclusion that the processes of oxygen accumulation/release by cerium oxide nanocrystals play an important role in the redox dynamics of nanoparticles when interacting with ROS can be made from the analysis of both our experimental data and data obtained by other authors. A red shift in the absorption spectra after adding hydrogen peroxide to a colloidal solution of cerium oxide nanocrystals observed in [3] is accompanied by a change in the color of the colloidal solution. This shift is due to the oxidation of Ce^{3+} to Ce^{4+} by means of a reaction that takes place on the surface of the nanoparticle: $Ce^{3+}+{}^{\cdot}OH \rightarrow Ce^{4+}+OH^{-}$. The reverse change in the color of the solution caused by the $Ce^{4+} \rightarrow Ce^{3+}$ transition lasted several tens of days, and the nature of this phenomenon in [3] did not find an explanation. This is the observation of the authors [3] can be explained by taking into account the processes of oxygen accumulation at the oxidation stage ($Ce^{3+} \rightarrow Ce^{4+}$) and oxygen release at the reduction stage ($Ce^{4+} \rightarrow Ce^{3+}$) of Ce^{3+} ions in cerium oxide nanocrystals.

As was already mentioned above, the content of Ce^{3+} ions in cerium oxide nanocrystals can reach 10–15%, while the doping by impurity ions can increase this content to 30–35%. Cerium ions are formed throughout the volume of the nanocrystal, but their concentration increases as they approach the surface.

The dynamics of the interaction of cerium oxide nanocrystals with ROS at the initial stage of interaction (Fig. 3.6) is characterized by high concentration of the oxidant, which leads to a significant number of acts of oxidation-reduction per unit of time. This opens up the possibility of observing a number of effects that are not

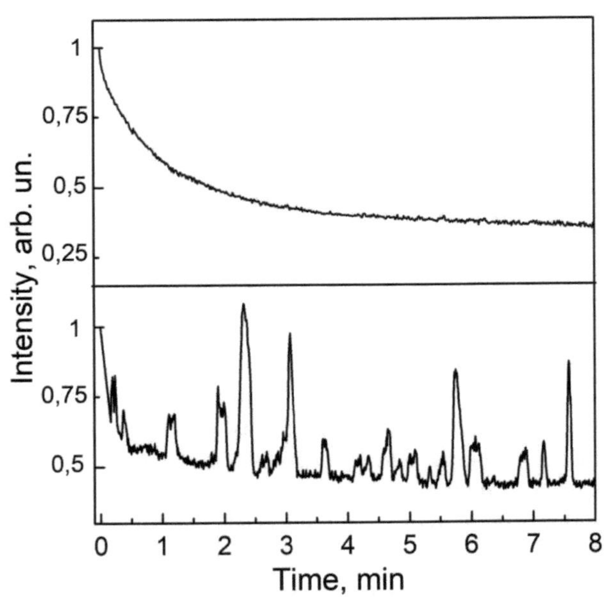

Fig. 3.6 Dynamics of changes in Ce^{3+} luminescence intensity at the initial stage of the process of interaction of CeO_2 nanocrystals (10 nm) with low (4 mM) and high (40 mM) concentrations of hydrogen peroxide

observed under normal conditions of interaction of reactive oxygen species with the surface of a nanoparticle (for example, in biological experiments, which are characterized by a rather low concentration of ROS).

The high initial ratio of Ce^{3+}/Ce^{4+} ions in cerium oxide nanocrystals on the one hand and the high concentration of the oxidant on the other hand lead to the fact that the "nanoparticle-oxidant" system can exhibit certain effects that are characteristic only for non-equilibrium systems. Such effects include chemical oscillations, in particular the well-known Belousov-Zhabotinsky oscillations, which are observed during the oxidation of citric acid with potassium bromate in the presence of Ce^{3+} ions, which act as a catalyst [59–61]. Chemical oscillations consist of periodic changes in the concentrations of reactants in a solution during a reaction, and their general interpretation is given within the framework of thermodynamics of non-equilibrium processes.

In our research, we observed an effect similar to Belousov-Zhabotinsky oscillations, namely the effect of long-term oscillations of the valence of cerium ions (Ce^{3+}/Ce^{4+}) in cerium oxide nanocrystals, which is observed immediately after the addition of large concentrations of hydrogen peroxide. This effect occurs at the stage immediately after the addition of hydrogen peroxide to aqueous solutions of nanocrystals and manifests itself in the occurrence of aperiodic oscillations in the intensity of Ce^{3+} luminescence, and the amplitude of oscillations can reach up to 50–60% of the initial intensity of luminescence (Fig. 3.7).

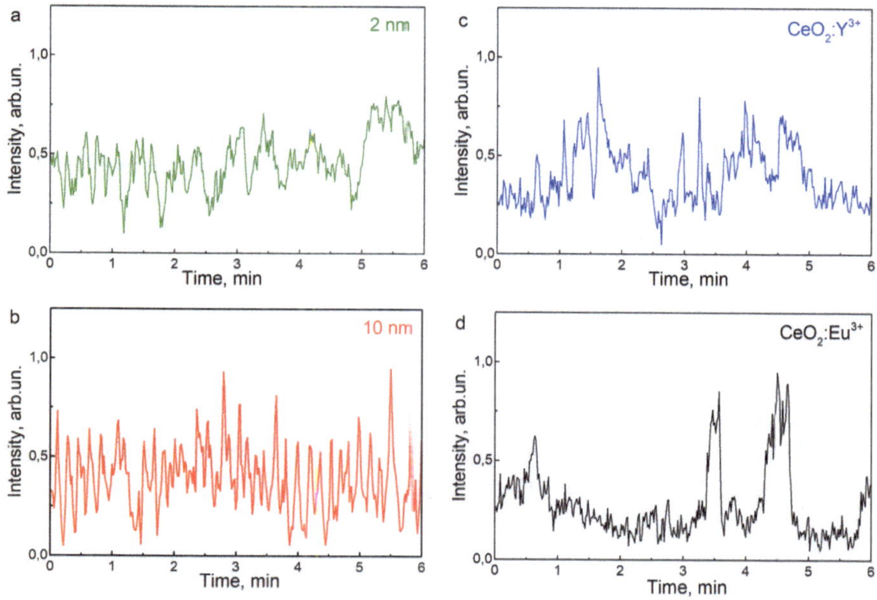

Fig. 3.7 Oscillations of Ce^{3+} luminescence intensity at hydrogen peroxide addition to colloidal solutions of cerium oxide nanocrystals (1—2 nm, 2—10 nm, 3—CeO_2-Y_2O_3 (50 nm), 4—CeO_2-Eu_2O_3 (50 nm))

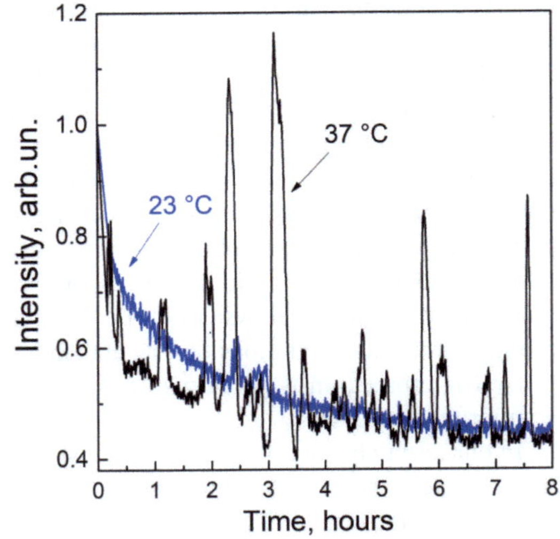

Fig. 3.8 Dynamics of changes in Ce^{3+} luminescence intensity upon addition of hydrogen peroxide ($C = 100$ mM) to colloidal solutions of CeO_{2-x} nanocrystals (2 nm) depending on temperature. At temperatures of 37 °C and higher, the transition of the system to the regime of oscillations of the Ce^{3+}/Ce^{4+} ion ratio in the nanocrystals is observed

It was shown that the oscillations are observed at concentrations of hydrogen peroxide exceeding 30–40 mM and at temperatures above 37 °C (Fig. 3.8).

These oscillations are in many aspects close to the Belousov-Zhabotinsky oscillations, in particular, in them the Ce^{3+}/Ce^{4+} ratio changes in the same way, but in this case, cerium ions are not in solution, but on the surface and in the volume of cerium oxide nanocrystals, and in the role hydrogen peroxide acting both as oxidant and reducing agent. This effect occurs only for nanoparticles with a high content of Ce^{3+} ions (from 5% and above). Thus, cerium valence oscillations were not observed for 50 nm cerium oxide nanoparticles with Ce^{3+} content up to 1.5%.

The oscillations in such a relatively simple system, consisting only of cerium oxide nanocrystals, hydrogen peroxide, and water, require additional explanations, and Sect. 3.2.1 will show a possible mechanism for the transition process during the interaction of cerium oxide nanocrystals with hydrogen peroxide, which leads to the formation if an oscillatory mode at high concentrations of hydrogen peroxide and Ce^{3+} ions.

3.2 Mechanisms of Antioxidant Activity of Cerium Oxide Nanocrystals During Interaction with ROS of Various Types (H_2O_2, ClO^-, O_2^-, $^{\cdot}OH$)

In order to clarify the microscopic mechanism of interaction of different types of ROS (H_2O_2, ClO^-, O_2^-, $^{\cdot}OH$) with the surface of cerium oxide nanocrystals and to determine the influence of the defect structure of nanocrystals (namely, the content of oxygen vacancies, Ce^{3+}/Ce^{4+} ions and Ce^{3+}-Vo-Ce^{3+} complexes on their antioxidant activity), a complex study of ROS decomposition processes on the surface of

CeO_{2-x} nanocrystals, as well as doped oxides was carried out using the methods of luminescence spectroscopy and optical absorption spectroscopy. For hydrogen peroxide and hypochlorite anions, the change in their concentration upon interaction with cerium oxide was determined by the change in the characteristic absorption bands at 250 nm (for hydrogen peroxide) [62] and 290 nm (for hypochlorite) [63]. Hydroxyl radicals and superoxide anions belong to the intermediate forms of ROS, so the dynamics of their formation and neutralization by CeO_{2-x} nanocrystals was determined indirectly by the oxidation of organic molecules, which at the same time are changed into a fluorescent form. In the work, coumarin, which turns into fluorescent 7-hydroxycoumarin upon interaction with $^{\cdot}OH$ radicals [64], was used to detect hydroxyl radicals, while adrenaline, which undergoes auto-oxidation with the participation of superoxide anions turning into fluorescent adrenochrome, was used to study the interaction of cerium oxide with superoxide anions [65].

3.2.1 *Interaction of Cerium Oxide Nanocrystals with Hydrogen Peroxide (H_2O_2)*

The ability of cerium oxide nanocrystals to effectively decompose hydrogen peroxide into molecular oxygen and water ($2\,H_2O_2 \rightarrow 2\,H_2O + O_2$) was shown in a number of publications [28, 50]. At the same time, the decomposition mechanisms proposed by different authors differ significantly. According to various models, the main role in the decomposition of hydrogen peroxide is played by Ce^{3+} (or Ce^{4+}) ions on the nanocrystal surface, Ce^{3+}-Vo-Ce^{3+} complexes, or Ce^{3+} ions formed in colloidal solutions of cerium oxide nanocrystals as a result of partial dissolution of nanoparticles. The key problem for all these concepts is that the interaction of hydrogen peroxide with individual Ce^{3+} ions should not lead to the decomposition of hydrogen peroxide into molecular oxygen and water ($2H_2O_2 \rightarrow 2H_2O + O_2$), but to the generation of much more reactive and dangerous for living cells hydroxyl- radicals ($^{\cdot}OH$) according to the Fenton mechanism [20]:

$$H_2O_2 + Ce^{3+} \rightarrow Ce^{4+} + OH^- + {}^{\cdot}OH \tag{3.5}$$

In order to bypass the possibility of formation of hydroxyl radicals ($^{\cdot}OH$) during the decomposition of H_2O_2, the authors of [50] proposed a model according to which hydrogen peroxide acts as a reducing agent when interacting with cerium oxide nanocrystals, therefore, the interaction process is accompanied by the $Ce^{4+} \rightarrow Ce^{3+}$ transition according to the reaction:

$$H_2O_2 + 2Ce^{4+} + 2OH^- \rightarrow 2H_2O + O_2 + 2Ce^{3+} \tag{3.6}$$

This model is not consistent with the results obtained in other publications, in which it was shown that hydrogen peroxide should act as an oxidizing agent and not a reducing agent in similar systems (at pH values typical for biological environments) [66]. Other authors suggested that formation of hydroxyl radicals ($^{\cdot}OH$) during the decomposition of hydrogen peroxide by nanocrystals of cerium oxide does occur, but the hydroxyl radicals formed at the initial stage of the process of decomposition of hydrogen peroxide at the next stage interact with hydrogen peroxide with the subsequent formation of superoxide anions (O_2^-) [39]:

$$^{\cdot}OH + H_2O_2 \rightarrow H_2O + H^+ + O_2^- \quad (3.7)$$

It should be noted, however, that both these models should be observed for cerium ions, which are not in nanoparticles, but rather in aqueous solutions of cerium salts, in which cerium is in ionic form. A similar model, as already mentioned above, was also proposed by a number of authors [42].

In order to show the difference (or similarity) in the interaction of cerium ions with hydrogen peroxide in cerium oxide nanocrystals and in aqueous solutions, aqueous solutions of CeO_2 nanocrystals and cerium chloride ($CeCl_3$), where cerium is in the ionic form (Ce^{3+}) and a comparison of the dynamics of changes in the concentration of hydrogen peroxide, the concentration of hydroxyl radicals, and the ratio of Ce^{3+}/Ce^{4+} ions in the process of decomposition of hydrogen peroxide was carried out.

Figure 3.9a shows the luminescence spectra of aqueous solutions of $CeCl_3$. The luminescence maximum of Ce^{3+} ions (350 nm) is noticeably shifted to the short-wavelength side in comparison with the luminescence maximum of Ce^{3+} ions in cerium oxide nanocrystals.

As in the case of CeO_{2-x} nanocrystals, the addition of hydrogen peroxide leads to decrease in the intensity of Ce^{3+} luminescence. At the same time, if the initial

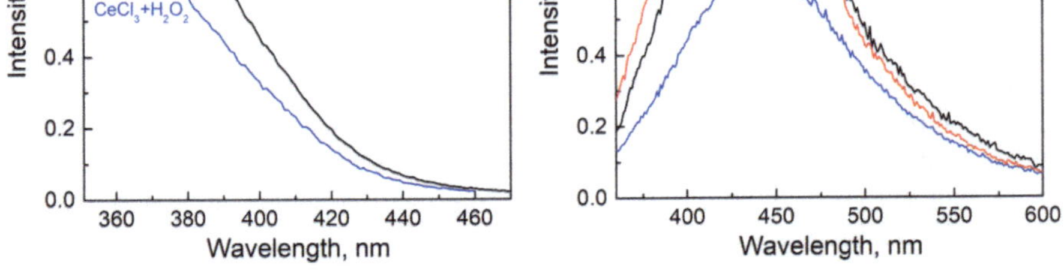

Fig. 3.9 Change in luminescence spectra during the interaction of individual Ce^{3+} ions (in aqueous $CeCl_3$ solution) (**a**) and CeO_{2-x} nanocrystals (**b**) with hydrogen peroxide. (Reprinted with permission from *Chemical Physics Letters*, vol. 767, V. Seminko, P. Maksimchuk, G. Grygorova, E. Okrushko, O. Sedyh, O. Avrunin, V. Semenets, Y. Malyukin, Switching the type of redox activity of colloidal nanoceria by Re^{3+} (Re = Y, Eu, Tb) doping, pp. 138363, 2021. Copyright 2021 Elsevier)

intensity of Ce^{3+} luminescence recovers over time for cerium oxide nanocrystals (the mechanism of this effect is described in Sect. 3.1), this effect was not observed for Ce^{3+} ions in the $CeCl_3$ solution (Fig. 3.10a).

Moreover, as expected, the decomposition of hydrogen peroxide by Ce^{3+} ions in the $CeCl_3$ solution occurs exclusively according to Fenton's mechanism, which is evidenced by a noticeable increase in the concentration of hydroxyl radicals in the process of decomposition of hydrogen peroxide in aqueous solution of cerium chloride (Fig. 3.10c). The change in the concentration of hydroxyl radicals in the process of decomposition of hydrogen peroxide was determined by the interaction of hydroxyl radicals with coumarin molecules. Coumarin does not have its own fluorescence, but it is converted to fluorescent 7-hydroxycoumarin (umbelliferone) during the reaction with hydroxyl radicals, thus, the fluorescence intensity of 7-hydroxycoumarin is proportional to the concentration of hydroxyl radicals in aqueous solutions.

At the same time, the decomposition of hydrogen peroxide by CeO_{2-x} nanocrystals do not provide additional hydroxyl radicals, but, on the contrary, the level of hydroxyl radicals when adding CeO_{2-x} nanocrystals turns out to be lower than in the control sample (aqueous solution of hydrogen peroxide without nanocrystals) (Fig. 3.10c). This indicates that individual Ce^{3+} ions cannot be responsible for the antioxidant activity of nanocrystals when interacting with hydrogen peroxide,

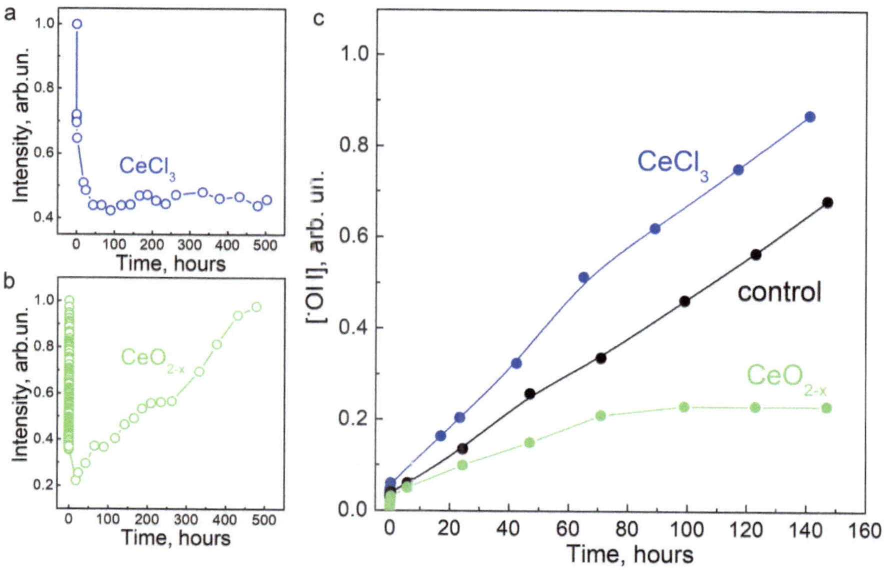

Fig. 3.10 Dynamics of changes in Ce^{3+} luminescence intensity after adding hydrogen peroxide to aqueous solutions of cerium chloride ($CeCl_3$) (**a**) and aqueous solutions of CeO_{2-x} nanocrystals (**b**); concentration of $^{\cdot}OH$ radicals in aqueous solutions of cerium chloride and CeO_{2-x} nanocrystals with H_2O_2 (**c**). (Reprinted with permission from *Chemical Physics Letters*, vol. 767, V. Seminko, P. Maksimchuk, G. Grygorova, E. Okrushko, O. Sedyh, O. Avrunin, V. Semenets, Y. Malyukin, Switching the type of redox activity of colloidal nanoceria by Re^{3+} (Re = Y, Eu, Tb) doping, pp. 138363, 2021. Copyright 2021 Elsevier)

because in this case, it should lead to the generation of excess of hydroxyl radicals ($^{\cdot}OH$) by the Fenton mechanism. The interaction of hydrogen peroxide with individual Ce^{3+} ions on the surface of cerium oxide nanocrystals should be accompanied by the generation of hydroxyl radicals, and the fact that this mechanism is not observed experimentally suggests that the antioxidant activity of cerium oxide nanocrystals cannot be caused by individual Ce^{3+} ions on the nanocrystal surface.

As can be seen from the results of research on the decomposition of hydrogen peroxide by CeO_{2-x} nanocrystals (the change in the concentration of hydrogen peroxide during the interaction of CeO_{2-x} nanocrystals with H_2O_2 was determined by the change in the intensity of the absorption band of hydrogen peroxide at 250 nm), the decrease in the concentration of hydrogen peroxide is accompanied by the oxidation of cerium ions ($Ce^{3+} \rightarrow Ce^{4+}$) (Fig. 3.11).

When the concentration of hydrogen peroxide becomes negligibly small, the reverse process begins ($Ce^{4+} \rightarrow Ce^{3+}$). As it was already shown in Sect. 3.1, this process is due to the release of oxygen from nanoparticle. In order to determine the mechanism of decomposition of hydrogen peroxide on the surface of CeO_{2-x} nanocrystals and the influence of the defect structure of the surface of nanocrystals on the processes of decomposition of hydrogen peroxide, the processes of interaction of hydrogen peroxide with cerium oxide nanocrystals of different sizes, as well as with nanocrystals doped by different ions, were studied.

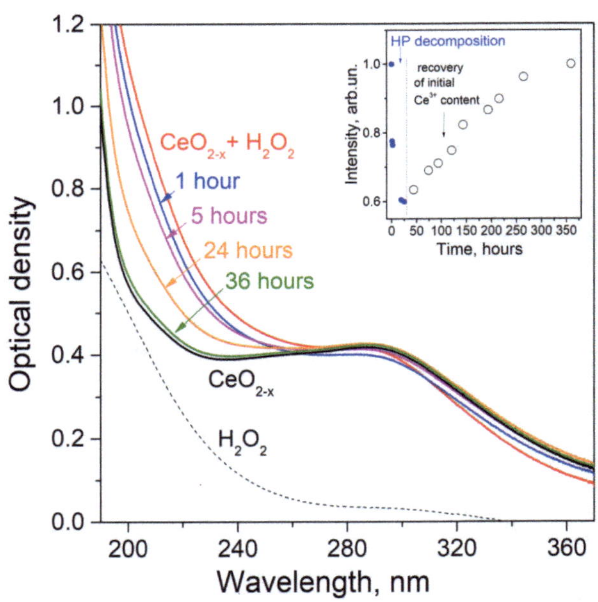

Fig. 3.11 Changes in absorption spectra and Ce^{3+} luminescence intensity during the interaction of cerium oxide nanocrystals with hydrogen peroxide. (Reprinted with permission from *Journal of Physical Chemistry C*, vol. 125, V. Seminko, P. Maksimchuk, G. Grygorova, E. Okrushko, O. Avrunin, V. Semenets, Y. Malyukin, Mechanism and dynamics of fast redox cycling in cerium oxide nanoparticles at high oxidant concentration, pp. 4743–4749, 2021. Copyright 2021 American Chemical Society)

Figure 3.12 shows the rate dependence of the hydrogen peroxide decomposition by cerium oxide nanocrystals. As expected, the decrease in the size of nanocrystals favors hydrogen peroxide decomposition processes. This result can be interpreted within the framework of any of the four concepts of the antioxidant activity of cerium oxide nanocrystals (a decrease in the size of nanocrystals leads to a higher content of oxygen vacancies (1), to an increase in the Ce^{3+}/Ce^{4+} ratio (2), to an increase in the content of Ce^{3+}-Vo-Ce^{3+} complexes (3), or to greater solubility of colloidal nanoparticles (4)).

We have observed that these dependences can be well described by well-known Michaelis-Menten kinetics [67]:

$$v = \frac{d[H_2O_2]}{dt} = \frac{V_{max}[H_2O_2]}{K + [H_2O_2]} \quad (3.8)$$

where K is the Michaelis constant, which is defined as the substrate concentration at which the decomposition rate is half the maximum; V_{max} is the maximum rate of decomposition of the substrate (hydrogen peroxide).

The fitting of the experimental data using this kinetics (Fig. 3.12, solid lines) allows to determine the catalytic parameters for both 2 nm and 10 nm nanoparticles including maximum rate of H_2O_2 decomposition V_{max} which is equal to 16.2 nmol/min for 2 nm and 6.5 nmol/min for 10 nm nanoparticles and Michaelis constant K (6.5 and 10 mM, respectively). Michaelis-Menten kinetics is an indicator of the presence of limited number of active catalytic sites in cerium oxide nanocrystals.

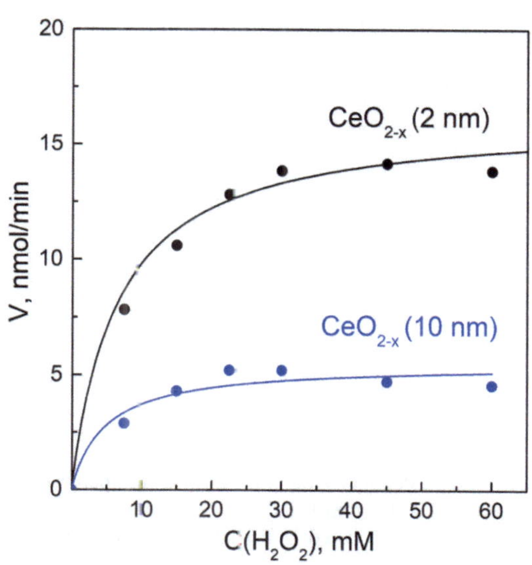

Fig. 3.12 Dynamics of hydrogen peroxide decomposition by cerium oxide nanocrystals of different sizes. (Reprinted with permission from *Journal of Physical Chemistry C*, vol. 125, V. Seminko, P. Maksimchuk, G. Grygorova, E. Okrushko, O. Avrunin, V. Semenets, Y. Malyukin, Mechanism and dynamics of fast redox cycling in cerium oxide nanoparticles at high oxidant concentration, pp. 4743–4749, 2021. Copyright 2021 American Chemical Society)

The nature of these sites as well as the mechanism of antioxidant action of cerium oxide nanocrystals was revealed by the study of hydrogen peroxide decomposition processes by doped nanocrystals.

If the ability of CeO_{2-x} nanocrystals to decompose hydrogen peroxide is really determined only by the Ce^{3+}/Ce^{4+} ratio on the surface of the nanocrystals, as suggested by the authors [2, 3], then an increase in the Ce^{3+}/Ce^{4+} ratio should lead to an increase in the rate of decomposition of hydrogen peroxide by nanocrystals. In this case, one could expect that the highest rate of hydrogen peroxide decomposition will be demonstrated by $CeO_2{:}Tb^{3+}$ nanocrystals, in which the content of Ce^{3+} ions is the highest according to the luminescence spectra (Fig. 3.13b), while the rate of hydrogen peroxide decomposition by CeO_{2-x} nanocrystals should be lower than for all nanocrystals of doped oxides. However, studies of the dependence of the decomposition time of hydrogen peroxide by CeO_{2-x} nanocrystals and nanocrystals of doped oxides on the initial concentration of hydrogen peroxide in the aqueous solution showed that the mechanism is actually somewhat different.

From Fig. 3.13a with the times of decomposition of hydrogen peroxide by doped nanocrystals determined from the absorption spectra for different initial concentrations of hydrogen peroxide, it is clear that for nanocrystals of cerium oxide with aliovalent impurity ions CeO_2-Re_2O_3 (Re = Y, Eu, Tb), the rate of hydrogen peroxide decomposition is lower than for CeO_{2-x} nanocrystals, while for Zr^{4+}-doped nanocrystals, an increase in the rate of hydrogen peroxide decomposition is observed in comparison with CeO_{2-x} nanocrystals (that is, the rate of decomposition of hydrogen

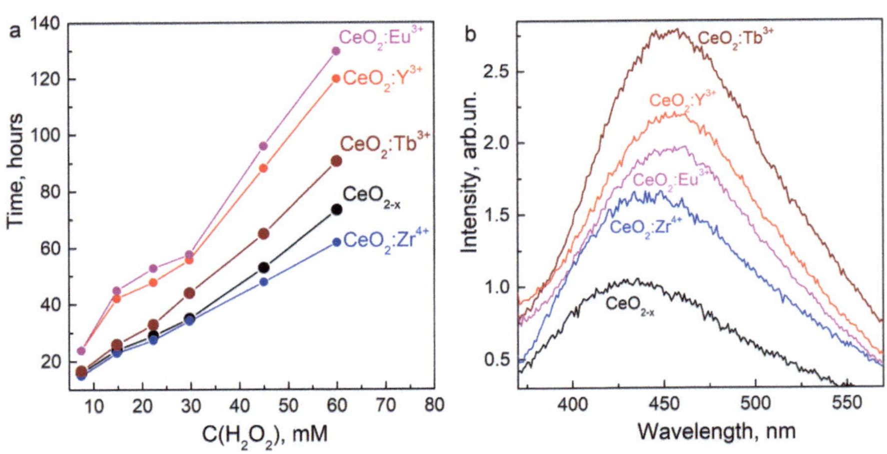

Fig. 3.13 (**a**) Dependence of the decomposition time of hydrogen peroxide by CeO_{2-x} nanocrystals and doped nanocrystals (C = 10 at.%) on the initial concentration of hydrogen peroxide in the aqueous solution; (**b**) luminescence spectra of colloidal solutions of CeO_{2-x} nanocrystals and doped oxide nanocrystals. For all studied samples: the size of nanocrystals is 2 nm, the concentration of doped ions is 10 at.%, C = 0.5 g/L. (Reprinted with permission from *Chemical Physics Letters*, vol. 767, V. Seminko, P. Maksimchuk, G. Grygorova, E. Okrushko, O. Sedyh, O. Avrunin, V. Semenets, Y. Malyukin, Switching the type of redox activity of colloidal nanoceria by Re^{3+} (Re = Y, Eu, Tb) doping, pp. 138363, 2021. Copyright 2021 Elsevier)

peroxide is not determined only by the content of Ce^{3+} ions on the surface of the nanocrystal, which contradicts the conclusions drawn by the authors [2, 3]).

This dependence cannot be explained if we assume that hydrogen peroxide interacts with individual Ce^{3+}/Ce^{4+} ions on the surface of nanocrystals, but it can be easily explained if we assume that the hydrogen peroxide molecule interacts not with individual Ce^{3+} ions, but with the Ce^{3+}-Vo-Ce^{3+} complex as a whole (Fig. 3.14). Similar ideas were previously proposed by the authors [49] in order to explain the role of oxygen vacancies in the formation of catalase- and SOD-like activity of cerium oxide nanocrystals, but they did not provide any experimental evidence. With this mechanism of interaction of hydrogen peroxide with the surface of nanocrystals, the oxidation of two Ce^{3+} ions to Ce^{4+} should occur at once with the splitting of the hydrogen peroxide molecule into two hydroxide anions (OH^-):

$$H_2O_2 + 2Ce^{3+} \rightarrow 2Ce^{4+} + 2OH^-, \tag{3.9}$$

which makes it possible to prevent the Fenton reaction (3.5), and, therefore, the formation of hydroxyl radicals, which would be inevitable during the interaction of a hydrogen peroxide molecule with individual Ce^{3+} ions on the surface of cerium oxide nanocrystals. Presence of Ce^{3+}-Vo-Ce^{3+} complexes on the surface of cerium oxide nanocrystals is thus one of the key factors providing the unique antioxidant effect of these nanocrystals.

Given the special role of Ce^{3+}-Vo-Ce^{3+} complexes in the formation of antioxidant properties of cerium oxide nanoparticles, it is easy to explain the results regarding the decrease in the decomposition rate of hydrogen peroxide for CeO_2-Re_2O_3 nanocrystals (Re = Y, Eu, Tb), and the acceleration of hydrogen peroxide decomposition processes for CeO_2-ZrO_2 nanocrystals. As was shown in Chap. 2, the content of oxygen vacancies in CeO_2-ZrO_2 nanocrystals is significantly higher than in CeO_{2-x} nanocrystals of the same size due to the weakening of oxygen bonds when introducing Zr^{4+} ions, which are much smaller in radius than Ce^{4+} ions. At the same time, the

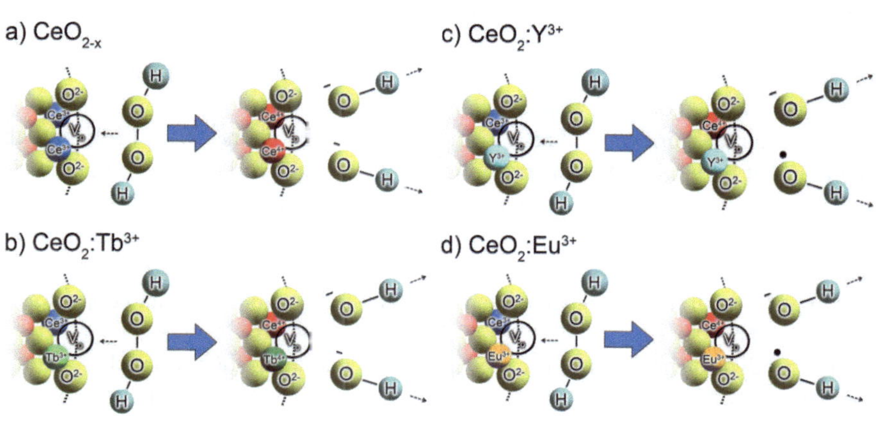

Fig. 3.14 Mechanisms of decomposition of hydrogen peroxide by CeO_{2-x} and doped CeO_2 nanoparticles. (Reprinted with permission from *Chemical Physics Letters*, vol. 767, V. Seminko, P. Maksimchuk, G. Grygorova, E. Okrushko, O. Sedyh, O. Avrunin, V. Semenets, Y. Malyukin, Switching the type of redox activity of colloidal nanoceria by Re^{3+} (Re = Y, Eu, Tb) doping, pp. 138363, 2021. Copyright 2021 Elsevier)

formation of oxygen vacancies is accompanied by the reduction of Ce^{4+} ions to Ce^{3+} ions, as a result of which the content of Ce^{3+}-Vo-Ce^{3+} complexes increases. A larger number of Ce^{3+}-Vo-Ce^{3+} complexes leads to the fact that the rate of decomposition of hydrogen peroxide by CeO_2-ZrO_2 nanocrystals is higher than that of CeO_{2-x} nanocrystals.

A decrease in the rate of hydrogen peroxide decomposition for CeO_2-Tb_2O_3 in comparison with CeO_{2-x} nanocrystals is not as significant as for CeO_2-Eu_2O_3 and CeO_2-Y_2O_3 nanocrystals. As was noted in Chap. 2, for CeO_2-Tb_2O_3 nanocrystals, the main type of defects at high concentrations of terbium (in particular, at $C = 10$ at.%, as in this experiment) are Tb^{3+}-Vo-Ce^{3+} complexes. Terbium ions can replace both Ce^{4+} and Ce^{3+} ions, so an increase in the concentration of terbium should lead to a decrease in the content of Ce^{3+}-Vo-Ce^{3+} complexes. But, on the other hand, the Tb^{3+}/Tb^{4+} ion is also an ion with a variable valence, so the Tb^{3+}-Vo-Ce^{3+} or Tb^{3+}-Vo-Tb^{3+} complexes can play a role similar to the role of Ce^{3+}-Vo-Ce^{3+} complexes in CeO_{2-x} nanocrystals during peroxide decomposition. The interaction of a hydrogen peroxide molecule with such a complex should lead to the oxidation of both Ce^{3+} ions to Ce^{4+} and Tb^{3+} to Tb^{4+} with the splitting of the hydrogen peroxide molecule into two hydroxide anions (OH^-):

$$H_2O_2 + 2Tb^{3+}\left(Ce^{3+}\right) \rightarrow 2Tb^{4+}\left(Ce^{4+}\right) + 2OH^- \quad (3.10)$$

A completely different picture should be observed for CeO_2-Eu_2O_3 and CeO_2-Y_2O_3 nanocrystals. As was shown in Chap. 2, for CeO_2-Eu_2O_3 and CeO_2-Y_2O_3 nanocrystals, the main type of defects at high concentrations of impurity ions (in particular, at impurity concentration of 10 at.%, as in this experiment) are Eu^{3+}-Vo-Ce^{3+} and Y^{3+}-Vo-Ce^{3+} complexes, respectively. At the same time, neither Eu^{3+} ion nor Y^{3+} ion can be oxidized to the tetravalent state, thus, the only possible mechanism for the interaction of a hydrogen peroxide molecule with the Eu^{3+}-Vo-Ce^{3+} (or Y^{3+}-Vo-Ce^{3+}) complex is the Fenton-like interaction: $H_2O_2 + Ce^{3+} \rightarrow Ce^{4+} + OH^- + {}^{\cdot}OH$. This mechanism, which was already mentioned above, is accompanied by the generation of hydroxyl radicals (${}^{\cdot}OH$), so CeO_2-Eu_2O_3 and CeO_2-Y_2O_3 nanocrystals, unlike CeO_{2-x}, CeO_2-Tb_2O_3 and CeO_2-ZrO_2 ones, should act not as antioxidants, but, on the contrary, as prooxidants.

This is a rather unexpected result, since today most authors [50, 52] adhere to the opinion that cerium oxide can act exclusively as an antioxidant (at least at physiological pH values). In order to test our hypothesis regarding the possible pro-oxidant role of CeO_2-Eu_2O_3 and CeO_2-Y_2O_3 nanocrystals, a study was conducted on the changes in the concentration of hydroxyl radicals (${}^{\cdot}OH$) in aqueous solutions during decomposition of hydrogen peroxide by cerium oxide nanocrystals and doped oxide nanocrystals.

The study has shown that, as expected from the assumptions we made regarding the features of the interaction of CeO_2-Eu_2O_3 and CeO_2-Y_2O_3 nanocrystals with hydrogen peroxide, the process of decomposition of hydrogen peroxide by these nanocrystals is accompanied by generation of hydroxyl radicals (${}^{\cdot}OH$) according to the Fenton mechanism (3.5) (Fig. 3.15).

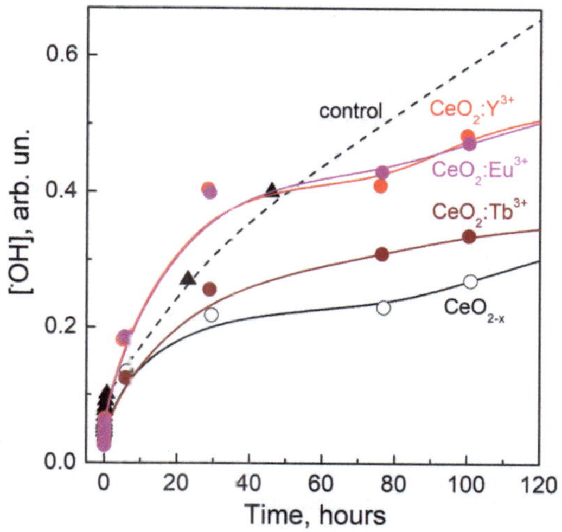

Fig. 3.15 Dynamics of changes in the concentration of hydroxyl radicals ($^{\cdot}OH$) in aqueous solutions during decomposition of hydrogen peroxide by CeO_{2-x}, CeO_2-Re_2O_3 (Re = Y, Eu, Tb) nanocrystals. (Reprinted with permission from *Chemical Physics Letters*, vol. 767, V. Seminko, P. Maksimchuk, G. Grygorova, E. Okrushko, O. Sedyh, O. Avrunin, V. Semenets, Y. Malyukin, Switching the type of redox activity of colloidal nanoceria by Re^{3+} (Re = Y, Eu, Tb) doping, pp. 138363, 2021. Copyright 2021 Elsevier)

Even more interesting was the decrease of the level of hydroxyl radicals which eventually became below control even for CeO_2-Eu_2O_3 and CeO_2-Y_2O_3 nanocrystals. This dynamics is significantly different from the dynamics of the interaction of hydrogen peroxide with individual Ce^{3+} ions in aqueous solutions of cerium chloride (Fig. 3.10c), where an increase in the concentration of hydroxyl radicals was observed throughout the process of decomposition of hydrogen peroxide.

Thus, based on the obtained results, it can be concluded that CeO_2-Eu_2O_3 and CeO_2-Y_2O_3 nanocrystals play a dual role in the processes of decomposition of hydrogen peroxide. On the one hand, these nanocrystals decompose hydrogen peroxide according to the Fenton mechanism ($H_2O_2 + Ce^{3+} \rightarrow Ce^{4+} + OH^- + {^{\cdot}OH}$), which is accompanied by the generation of hydroxyl radicals ($^{\cdot}OH$). But, on the other hand, the same nanocrystals are able to scavenge hydroxyl radicals formed in the process of decomposition of hydrogen peroxide (details of this mechanism are discussed in Sect. 3.2.3). Neutralization of hydroxyl radicals (reducing them to hydroxide anions) occurs by oxidation of individual Ce^{3+} ions on the surface of nanoparticles. Since surface Ce^{3+} ions are involved both in the processes of decomposition of hydrogen peroxide and in the processes of neutralization of hydroxyl radicals, the decomposition of hydrogen peroxide by these nanocrystals occurs more slowly than for CeO_{2-x}, CeO_2-Tb_2O_3 and CeO_2-ZrO_2 nanocrystals.

So, Ce^{3+}-Vo-Ce^{3+} complexes really play the main role in the processes of decomposition of hydrogen peroxide on the surface of cerium oxide nanocrystals. The role of these complexes should be similar to the role of the active centers of enzymes, which makes it possible to use the mathematical apparatus that was previously used to describe the kinetics of enzyme action. Further denoting the concentration of hydrogen peroxide by $[H_2O_2]$, and taking into account that the number of Ce^{3+}-Vo-Ce^{3+} complexes on the surface of nanocrystals is much smaller than the number of hydrogen peroxide molecules, one can obtain the dependence of the rate of formation of the reaction product (hydroxide anions) on the concentration of hydrogen peroxide (this rate, in turn, is equal to the rate of decomposition of hydrogen peroxide).

$$v = \frac{d\left[H_2O_2\right]}{dt} = \frac{V_{max}\left[H_2O_2\right]}{K + \left[H_2O_2\right]} \tag{3.11}$$

This kinetics is completely analogous to the well-known Michaelis-Menten kinetics [67], where K is the Michaelis constant, which is defined as the substrate concentration at which the decomposition rate is half the maximum; V_{max} is the maximum rate of decomposition of the substrate (hydrogen peroxide).

In order to check the adequacy of this mathematical model with experimental data, the dependences of the decomposition time of hydrogen peroxide on the initial peroxide concentration (Fig. 3.13a) were transformed to dependences of the decomposition rate of hydrogen peroxide on the initial concentrations of peroxide (Fig. 3.16). From these dependences, it is clear that an increase in the initial concentration of hydrogen peroxide at low peroxide concentrations (up to 30 mM) leads to

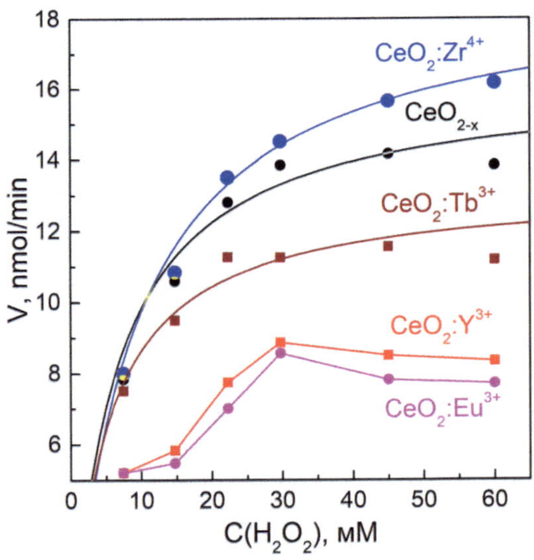

Fig. 3.16 Dependence of the rate of decomposition of hydrogen peroxide by doped nanocrystals on the initial concentration of hydrogen peroxide in aqueous solutions. (Reprinted with permission from *Chemical Physics Letters*, vol. 767, V. Seminko, P. Maksimchuk, G. Grygorova, E. Okrushko, O. Sedyh, O. Avrunin, V. Semenets, Y. Malyukin, Switching the type of redox activity of colloidal nanoceria by Re^{3+} (Re = Y, Eu, Tb) doping, pp. 138363, 2021. Copyright 2021 Elsevier)

an increase in the rate of decomposition of hydrogen peroxide, since an increase in the concentration of peroxide under these conditions leads to an increase in the quota of Ce^{3+}-Vo-Ce^{3+} complexes involved in decomposition of hydrogen peroxide.

At hydrogen peroxide concentrations higher than 30 mM, the growth rate with concentration slows down and then reaches saturation, i.e., at these concentrations, all or almost all Ce^{3+}-Vo-Ce^{3+} complexes on the surface of the nanocrystal are involved in the decomposition of hydrogen peroxide.

For CeO_{2-x}, CeO_2-Tb_2O_3 and CeO_2-ZrO_2 nanocrystals dependence of the rate of decomposition of hydrogen peroxide by nanocrystals on the initial concentration of hydrogen peroxide is well described within the framework of the proposed mathematical model. For CeO_2-Eu_2O_3 and CeO_2-Y_2O_3 nanocrystals, the approximation by formula (3.11) leads to unsatisfactory results, that is explained by the fact that during the decomposition of hydrogen peroxide, Re^{3+}-Vo-Ce^{3+} (Re = Y, Eu) complexes on the surface of CeO_2-Eu_2O_3 and and CeO_2-Y_2O_3 nanocrystals interact not with one (that was assumed in formula 3.13), but with two substrates (hydrogen peroxide and hydroxyl radicals), which leads to competition between substrates for an active center (Re^{3+}-Vo-Ce^{3+} complex).

Drawing an analogy between the catalytic action of enzymes and the catalytic action of cerium oxide nanocrystals, and using the approximation of experimental results by formula 3.11, it is possible to determine the catalytic parameters of CeO_{2-x}, CeO_2-Tb_2O_3 and CeO_2-ZrO_2 nanoparticles, which are given in Table 3.1.

A more detailed consideration of the direction and mechanisms of redox processes in the Ce^{3+}/Ce^{4+} + H_2O_2 system can be carried out using the Pourbaix diagrams known from electrochemistry, which are used to determine the direction of redox processes in chemical systems depending on redox potentials and pH [68].

Pourbaix diagrams are calculated on the basis of the Nernst equation, which for the Ce^{3+}/Ce^{4+} ion pair has the form:

$$E(\text{eV}) = 1.44 - 0.059\log\left[Ce^{3+}\right]/\left[Ce^{4+}\right] \tag{3.12}$$

From the Pourbaix diagram for the Ce^{3+}/Ce^{4+} ion pair (Fig. 3.17), on which the straight lines corresponding to hydrogen peroxide as an oxidizing agent ($H_2O_2 + 2e^- \rightarrow 2OH^-$) and as a reducing agent ($H_2O_2 \rightarrow O_2 + 2H^+ + 2e^-$) are shown, it can be seen that at initial pH = 7.5, hydrogen peroxide acts exclusively as an

Table 3.1 Catalytic parameters of CeO_{2-x}, CeO_2-Tb_2O_3 and CeO_2-ZrO_2 nanoparticles (V_{max}—maximum rate of hydrogen peroxide decomposition by nanoparticles, K—Michaelis constant (concentration of hydrogen peroxide at which the rate of decomposition is half the maximum), k_{cat}—rate of decomposition of hydrogen peroxide by one nanoparticle, k_{cat}/K is the catalytic efficiency of the nanoparticle)

	CeO_2-ZrO_2	CeO_{2-x}	CeO_2-Tb_2O_3
V_{max}, nmol/min	19	16.2	13.2
K, mM	9.5	6.5	5.3
k_{cat}, molecules/NP/min	11	9.5	7.7
k_{cat}/K, M^{-1} s^{-1}	24.4	19.3	24.2

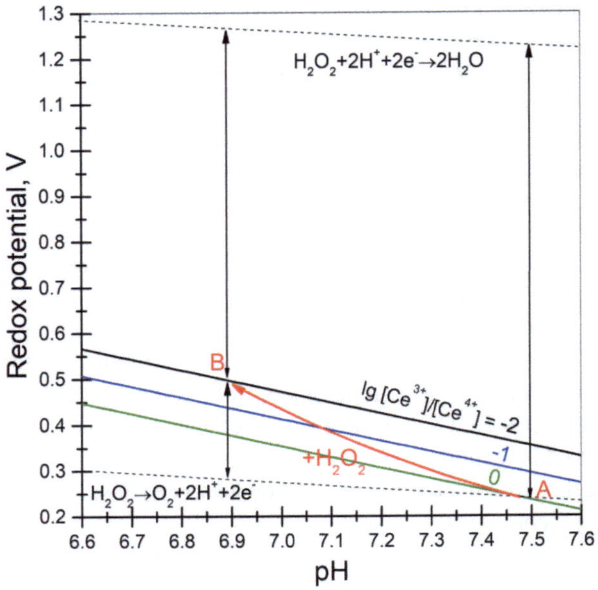

Fig. 3.17 Pourbaix diagram for the system Ce^{3+}/Ce^{4+} + H_2O_2. (Reprinted with permission from *Journal of Physical Chemistry C*, vol. 125, V. Seminko, P. Maksimchuk, G. Grygorova, E. Okrushko, O. Avrunin, V. Semenets, Y. Malyukin, Mechanism and dynamics of fast redox cycling in cerium oxide nanoparticles at high oxidant concentration, pp. 4743–4749, 2021. Copyright 2021 American Chemical Society)

oxidant and, therefore, a significant part of Ce^{3+} ions must turn into Ce^{4+} to achieve electrochemical equilibrium in the system. At the same time, a decrease in pH and the transition of a part of Ce^{3+} ions to Ce^{4+} can lead to the fact that hydrogen peroxide begins to show its reducing properties, which, as will be shown below, can play a key role, in particular, in the occurrence of the effect of Ce^{3+}/Ce^{4+} valence oscillations when adding significant concentrations of hydrogen peroxide to aqueous solutions of cerium oxide nanoparticles.

Using the Pourbaix diagram, it can be concluded that the percentage of Ce^{3+} ions that change to Ce^{4+} when hydrogen peroxide is added should depend both on the initial pH of the solution and on the initial ratio of Ce^{3+}/Ce^{4+} ions in the nanoparticle. Indeed, the decrease of nanoparticle size from 50 nm to 2 nm (Fig. 3.18a) and the addition of sodium hydroxide (which has pronounced alkaline properties) to an aqueous solution (Fig. 3.18b) lead to an increase of the part of Ce^{3+} ions that change to Ce^{4+} when hydrogen peroxide is added. The increase in the percentage of Ce^{3+} ions that change to Ce^{4+} when hydrogen peroxide is added in these cases is due to the increase in the content of Ce^{3+} ions during the transition to small nanoparticles and the increase in the initial pH when sodium hydroxide is added. It can be seen from Pourbaix diagram (Fig. 3.17) that both should lead to the stabilization of cerium ions in the tetravalent state.

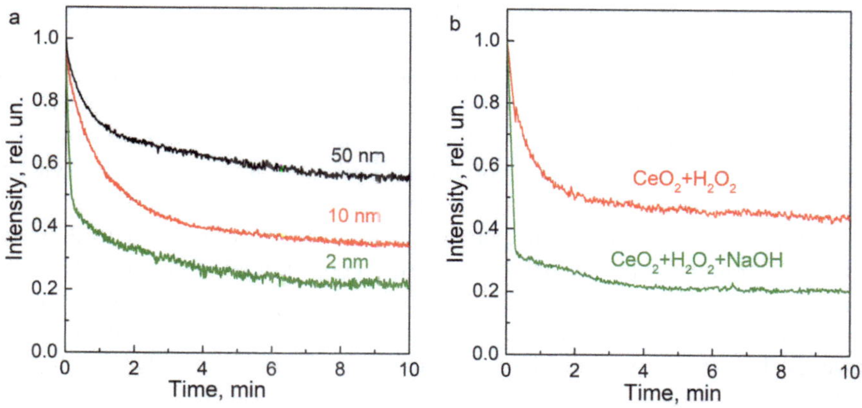

Fig. 3.18 Dynamics of changes in the intensity of Ce^{3+} luminescence when hydrogen peroxide (50 mM) is added to colloidal solutions of CeO_{2-x} nanocrystals of different sizes and the pH of the solution changes

At the same time, it is important to note that equilibrium (which means the transition of the corresponding part of Ce^{3+} ions to Ce^{4+}) is not achieved by this system. The very fact that this system is non-equilibrium one leads to the possibility of observing the effect of long-term oscillations of the intensity of Ce^{3+} luminescence after the addition of high concentrations of hydrogen peroxide, which were described in Sect. 3.1.

As already mentioned, these oscillations are in many aspects close to the well-known Belousov-Zhabotinsky oscillations observed during the oxidation of citric acid with potassium bromate in the presence of Ce^{3+} ions, which act as a catalyst in the system. However, in our case, cerium ions are on the surface and in the volume of nanocrystals, and not in solution, while hydrogen peroxide acts as both an oxidizing and reducing agent. As already mentioned, oscillations were observed only for concentrations above 40 mM, which is close to the concentrations at which the Michaelis-Menten kinetics for these nanoparticles reaches saturation, i.e., when almost all catalytic sites on the surface of the nanoparticle are involved.

The analysis of Pourbaix diagram (Fig. 3.17) opens the way to understand the general mechanism of occurrence and synchronization of Ce^{3+}/Ce^{4+} oscillations in nanoceria. The initial pH value of nanoceria water solution is 7.5, for which the redox potentials of H_2O_2/H_2O redox couple are ~1.2 V and ~0.25 V for O_2/H_2O_2 redox couple. Before H_2O_2 addition, the concentrations of Ce^{3+} and Ce^{4+} ions on the surface of ceria nanoparticle can be considered to be close to equal leading to redox potential of Ce^{3+}/Ce^{4+} redox couple of ~0.23 V (or even less if the number of Ce^{3+} ions is higher than Ce^{4+} ones). So, at pH 7.5, the value of Ce^{3+}/Ce^{4+} redox potential (point A on the Pourbaix diagram) is lower than redox potential of H_2O_2/H_2O redox couple, and H_2O_2 should act as an oxidant. In other words, at pH 7.5 H_2O_2 decomposition by nanoceria via $H_2O_2 + 2H^+ + 2e^- \rightarrow 2H_2O$ route takes place. At the same time, Ce^{4+} ions at pH 7.5 cannot oxidize H_2O_2 to oxygen as the redox potential of

Ce^{3+}/Ce^{4+} redox couple at high Ce^{3+}/Ce^{4+} ratio is less than the one of O_2/H_2O_2 redox couple.

Instant addition of high H_2O_2 concentrations drastically changes the redox balance in the system. In Fig. 3.19, the change of Ce^{3+} luminescence intensity and change of pH of nanoceria colloidal solutions measured few minutes after H_2O_2 addition are shown on the single plot. Ce^{3+}/Ce^{4+} ratio and pH decrease synchronously (both concentration dependences are linear on the half-log scale).

Further increase of H_2O_2 concentration (more than 100 mM) do not lead to additional Ce^{3+} luminescence quenching, so at such concentrations one can suppose that luminescence of all near-surface Ce^{3+} ions is fully quenched due to oxidation to Ce^{4+} ones and remaining luminescence belong to those bulk Ce^{3+} ions which are directly inaccessible to the oxidant. The oscillations are observed immediately after this synchronous pH and Ce^{3+}/Ce^{4+} ratio drop caused by initial H_2O_2 addition, so we should admit that this instant drop actually triggers the long-lasting Ce^{3+}/Ce^{4+} redox cycling. The mechanism leading to Ce^{3+}/Ce^{4+} oscillations is the following. Addition of high H_2O_2 concentrations leads to instant pH decrease from 7.5 to 6.9–7.1. At the same time, the most part of Ce^{3+} ions are oxidized to Ce^{4+} ones. As a result, value of redox potential of Ce^{3+}/Ce^{4+} redox couple (point B on the Pourbaix diagram) becomes higher than the value of redox potential of O_2/H_2O_2 redox couple. Most of cerium ions are now in Ce^{4+} state and according to Pourbaix diagram, they can oxidize H_2O_2 leading to H_2O_2 decomposition by an alternative $H_2O_2 \rightarrow O_2 + 2H^{+} + 2e^{-}$ route.

Synchronous reduction of Ce^{4+} ions to Ce^{3+} leads to equally synchronous oxidation of Ce^{3+} ions at the next stage, and so on. Notably, for 2 nm nanoparticles, the frequency of oscillations (about 10 oscillations/min) is surprisingly close to the maximum rate of H_2O_2 decomposition by single nanoparticle which can be

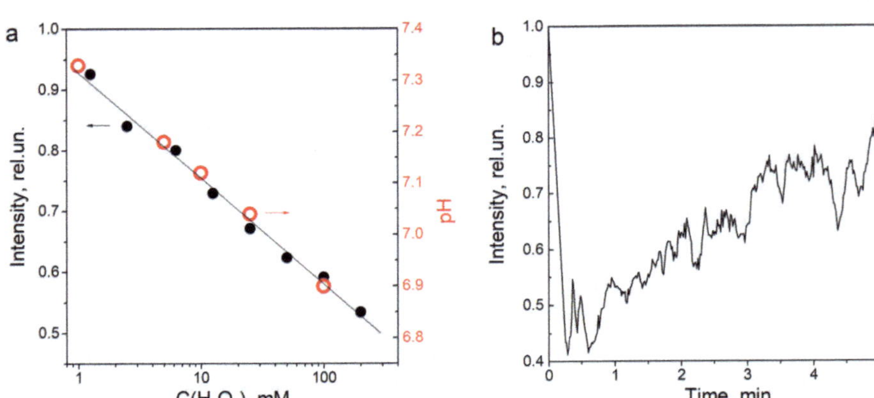

Fig. 3.19 (**a**) Ce^{3+} luminescence intensity and pH of 2 nm nanoceria water solutions 10 min after addition of different H_2O_2 concentrations. (**b**) pH-induced oscillations in 2 nm nanoceria at low (10 mM) H_2O_2 concentrations. (Reprinted with permission from *Journal of Physical Chemistry C*, vol. 125, V. Seminko, P. Maksimchuk, G. Grygorova, E. Okrushko, O. Avrunin, V. Semenets, Y. Malyukin, Mechanism and dynamics of fast redox cycling in cerium oxide nanoparticles at high oxidant concentration, pp. 4743–4749, 2021. Copyright 2021 American Chemical Society)

determined from Michaelis-Menten kinetics (16.2 nmol/min or ~9.5 molecules/min per one nanoparticle). Oscillation frequency for 10 nm nanoparticles is about two times less than for 2 nm ones (5–6 oscillations/min), as the rate of H_2O_2 decomposition by 10 nm nanoparticles is lower as well. Additional synchronization of the oscillations of individual nanocrystals in an aqueous solution is achieved by the fact that the decomposition of hydrogen peroxide using one or another mechanism is accompanied by corresponding pH change (an increase in pH during $Ce^{3+} \rightarrow Ce^{4+}$ transition and decrease in pH during the reverse transition), which, as a result, synchronizes the dynamics of changes of cerium valence in nanocrystals within the solution.

This role of pH is confirmed by an experiment in which hydrogen peroxide was added to an aqueous solution of nanocrystals with a concentration insufficient to cause oscillations. It was found that if at the same time, the pH of the solution is quickly increased, for example, by the simultaneous addition of a high concentration of sodium hydroxide (NaOH), then oscillations appear in such a system (Fig. 3.19b).

Thus, the key role of Ce^{3-}-Vo-Ce^{3-} complexes in the processes of interaction of cerium oxide nanocrystals with hydrogen peroxide was revealed, and it was shown that the very fact that Ce^{3+} ions are mainly in the form of such complexes provides an anti- rather than a prooxidant effect during the interaction with H_2O_2 molecules.

3.2.2 *Interaction of Cerium Oxide Nanocrystals with Hypochlorite Anions (ClO^-)*

Hypochlorites (primarily sodium hypochlorite NaOCl and calcium hypochlorite $Ca(OCl)_2$) are widely known as highly effective bleaching and disinfecting agents. Recently, their important biological role was shown [69]. The immune activity of neutrophils and monocytes is accompanied by so-called respiratory burst, which consists in the extremely rapid consumption of oxygen by cells and the subsequent generation of superoxide anions (O_2^-) and hydrogen peroxide (H_2O_2), which play a key role in the destruction of infectious agents [70–72]. Recent studies in this field have shown that superoxide and hydrogen peroxide do not directly participate in the destruction of pathogens, since their microbicidal activity is quite limited, but they can interact with chlorine ions (this reaction is catalyzed by myeloperoxidase) with the formation of hypochlorite anions (ClO^-), which are much more active in killing pathogens [73]. Hypochlorite anions are strong nonselective oxidants that can interact with DNA [74] and lipid membranes [75], oxidize amino acids cysteine and methionine [76], and interact with amino groups to form chloramines, which, in turn, can oxidize thiols, thioethers, and heme proteins [77], which eventually lead to the destruction of the pathogenic cell. However, just as they destroy infectious agents, hypochlorite anions can also damage healthy body cells. In addition, hypochlorous acid (HOCl) and hypochlorite anions (ClO^-) are believed to provoke inflammatory diseases causing respiratory distress syndrome, ischemic lesions, and arthritis [78].

The study of the mechanisms of protection of cells against the effects of hypochlorite anions has shown that cells lack any enzyme-based defense system for the removal of hypochlorite anions, similar to the systems used to control the concentrations of superoxide and hydrogen peroxide (superoxide dismutase and catalase, respectively). Therefore, the increase in the concentration of hypochlorite anions leads to the oxidation of intracellular glutathione, which can be replaced only by resynthesis. Thus, hypochlorite-specific artificial antioxidants, in particular, cerium oxide nanocrystals, can play a particularly important role in controlling the concentration of hypochlorite anions. To date, there are only a few publications devoted to the interaction of cerium oxide nanocrystals with hypochlorite anions [79], and none in which the processes of the interaction of hypochlorite anions with cerium oxide nanocrystals in dynamics were studied in detail. By combining the methods of optical absorption spectroscopy and luminescence spectroscopy, we conducted a comprehensive study of the mechanisms and dynamics of the interaction of hypochlorite anions with cerium oxide nanocrystals at varying hypochlorite concentrations.

The concentrations of cerium oxide nanocrystals in aqueous solutions were the same in all experiments and equal to 0.1 g/L. In the experiments, 30 μL of aqueous solutions of sodium hypochlorite with a hypochlorite concentration of 0.5, 1, 2, and 4% were added to 30 ml of aqueous colloidal solutions of cerium oxide nanocrystals. Thus, the concentration of hypochlorite in the solutions was 2 mM, 4 mM, 8 mM, and 16 mM, respectively.

The processes of decomposition of hypochlorite anions by cerium oxide nanoparticles were accompanied by a change in the intensity of the absorption band of hypochlorite anions at 290 nm. The hypochlorite solution without nanoparticles was stable and its storage at room temperature for several weeks did not lead to any changes in the intensity of the absorption band at 290 nm. Also, no changes in the intensity of the 290 nm band during 10 days were observed when hypochlorite was added to solutions containing nanoparticles other than cerium oxide (in particular, $CePO_4$). In turn, the presence of cerium oxide nanoparticles in the solution significantly increases the rate of decomposition of hypochlorite making it noticeable even at room temperature (22 °C). Figure 3.20 shows the absorption spectra of aqueous colloidal solutions of 2 nm cerium oxide nanoparticles ($C = 0.1$ g/L) before and after the addition of different concentrations of NaOCl. The intensity of the band at 290 nm decreases with time, which indicates the gradual decomposition of hypochlorite by cerium oxide nanoparticles. The time required for the decomposition of hypochlorite is proportional to the initial concentration and is the least for 2 mM and the highest for 16 mM solutions.

The process of decomposition of hypochlorite was accompanied by the formation of oxygen. The concentration of dissolved oxygen, measured using the Winkler method after the end of the hypochlorite decomposition process, was close to the maximum oxygen concentration in water at a temperature of 22 °C (40 mg/L) for all samples. It is interesting that the decomposition of hypochlorite does not lead to a complete recovery of the initial absorption spectra of cerium oxide, and the overall intensity of the absorption band with charge transfer (Ce^{4+}-O^{2-}) after the decomposition of hypochlorite was higher than before the addition of hypochlorite. This effect is probably related to the modification of the cerium oxide surface upon addition of hypochlorite (for example, the chlorination of the cerium oxide surface proposed in [79]).

Time dependences of hypochlorite concentration in aqueous solutions with cerium oxide nanoparticles, determined by absorption spectra, are shown in the insets of Fig. 3.20. For all concentrations of hypochlorite, a similar behavior of concentration with time was observed, namely, a fast decomposition of

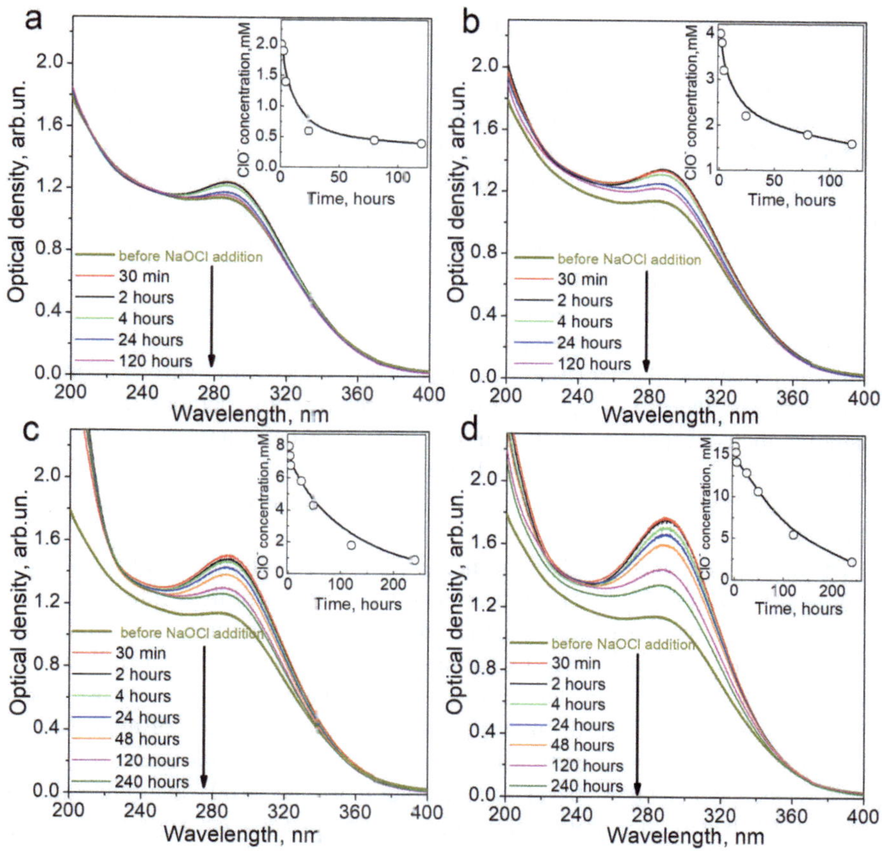

Fig. 3.20 Absorption spectra before and after adding different concentrations of NaOCl (**a**—2 mM, **b**—4 mM, **c**—8 mM, **d**—16 mM) to aqueous colloidal solutions of 2 nm CeO_{2-x} nanocrystals (0.1 g/L). The insets show the time dependences of the NaOCl concentration in aqueous solutions obtained from the change in the intensity of the absorption band at 290 nm. (Reprinted with permission from *Journal of Physical Chemistry C*, vol. 123, V. Seminko, P. Maksimchuk, G. Grygorova, O. Avrunin, V. Semenets, V. K. Klochkov, Yu.V. Malyukin, Catalytic Decomposition of Hypochlorite Anions by Ceria Nanoparticles Visualized by Spectroscopic Techniques, pp. 20675–20681, 2029. Copyright 2019 American Chemical Society)

hypochlorite at the initial stage with a gradual slowdown, which occurs, most likely, due to a decrease in the concentration of hypochlorite, which leads to a decrease in the probability of interaction of the hypochlorite molecule with the catalytic site on the surface of the nanocrystal.

To date, literature lacks a detailed interpretation of the mechanisms of the interaction of cerium oxide nanocrystals with ClO^- anions, which lead to the decomposition of hypochlorite. In work [79], in which the authors investigated the change in the oxidation state of cerium during the decomposition of hypochlorite by cerium oxide nanocrystals using the methods of X-ray photoelectron spectroscopy (XPS) and optical absorption spectroscopy, the following reaction between Ce^{4+} ions and hypochlorite anion was proposed:

$$Ce^{4+} + 2ClO^- \rightarrow Ce^{3+} + 2Cl^- + O_2 \tag{3.13}$$

Therefore, according to their hypothesis, hypochlorite should reduce Ce^{4+} to Ce^{3+}, which is rather surprising, given that hypochlorite is usually used to oxidize Ce^{3+} ions to Ce^{4+} in aqueous solutions, and not the other way. Also, according to the Nernst Eq. (3.12), the standard potential of the Ce^{4+}/Ce^{3+} pair under neutral and reducing conditions (initial pH in our experiments was 7.5) is lower than the corresponding value for NaOCl (0.89 V) [68], which confirms that NaOCl should play the role of an oxidizing agent.

To understand the details of the interaction of CeO_{2-x} nanocrystals with hypochlorite anions, a complex approach was used, which was applied in Sect. 3.2.1 to study the processes of interaction of CeO_{2-x} nanocrystals with hydrogen peroxide, which involved the use of optical absorption spectroscopy methods (to study the change in the concentration of ROS in the process of their interaction with CeO_{2-x} nanocrystals) as well as methods of luminescence spectroscopy to determine the change in the ratio of Ce^{3+}/Ce^{4+} ions in cerium oxide nanocrystals during interaction with ROS.

The time dependences of Ce^{3+} luminescence intensity after the addition of hypochlorite are shown in Fig. 3.21 for different concentrations of hypochlorite.

It can be seen that the decrease in the intensity of Ce^{3+} luminescence, which corresponds to the dominant role of oxidation ($Ce^{3+} \rightarrow Ce^{4+}$) processes, is followed by the stages of slow and then fast recovery of the intensity of Ce^{3+} luminescence (Fig. 3.21), which are well distinguished at a high concentration of hypochlorite. The complete redox cycle ends with the recovery of the initial intensity of Ce^{3+} luminescence, which takes 50 h at a concentration of 2 mM, 100 h at a concentration of 4 mM, 175 h at a concentration of 8 mM, and 285 h at a concentration of 16 mM. It is worth noting that the dynamics of recovery of Ce^{3+} luminescence intensity at the last stage is almost independent of hypochlorite concentration. The fraction of Ce^{3+} ions that remained unoxidized after the end of the oxidation stage also depends on the hypochlorite concentration and is equal to 0.4 for 2 mM, 0.3 for 4 mM, 0.22 for 8 mM, and 0.13 for 16 mM NaOCl.

Studies of changes in hypochlorite concentration (Fig. 3.20) and intensity of Ce^{3+} luminescence (Fig. 3.21) allow us to propose a mechanism of interaction between

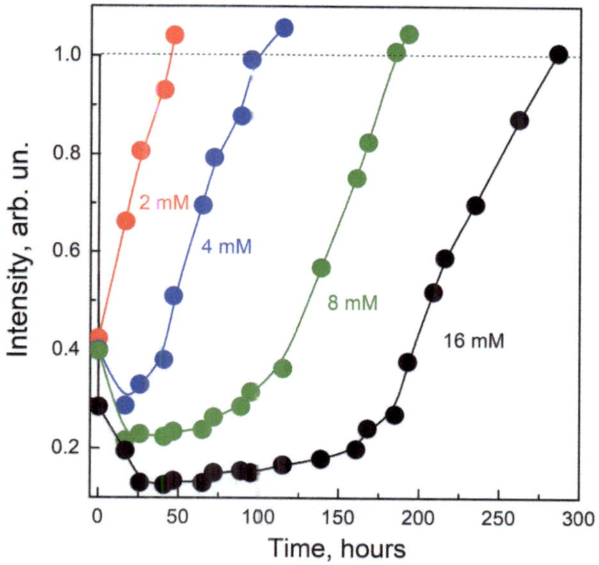

Fig. 3.21 Change in intensity of Ce^{3+} luminescence of cerium oxide nanocrystals after adding different concentrations of NaOCl. (Reprinted with permission from *Journal of Physical Chemistry C*, vol. 123, V. Seminko, P. Maksimchuk, G. Grygorova, O. Avrunin, V. Semenets, V. K. Klochkov, Yu.V. Malyukin, Catalytic Decomposition of Hypochlorite Anions by Ceria Nanoparticles Visualized by Spectroscopic Techniques, pp. 20675–20681, 2029. Copyright 2019 American Chemical Society)

cerium oxide nanocrystals and hypochlorite anions. The main role in the decomposition of hypochlorite anions belongs to the Ce^{3+}-Vo-Ce^{3+} complexes, which play the role of catalytic centers. According to the Nernst equation and Pourbaix diagrams, hypochlorite can only act as an oxidizing agent at pH = 7.5. At the same time, the reaction of oxidation of Ce^{3+} ions with hypochlorite looks like this:

$$ClO^- + H_2O + 2e^- \rightarrow Cl^- + 2OH^- \tag{3.14}$$

This reaction should be accompanied by the formation of additional hydroxide ions (OH^-) and, thus, an increase in pH. However, the addition of cerium oxide nanoparticles to aqueous hypochlorite solutions did not result in any observable pH changes. Therefore, it can be assumed that after the oxidation occurring on the catalytic sites Ce^{3+}-Vo-Ce^{3+}:

$$ClO^- + 2Ce^{3+} \rightarrow 2Ce^{4+} + O^{2-} + Cl^- \tag{3.15}$$

oxygen ions (O^{2-}) do not return to the aqueous solution with the formation of hydroxide anions OH^-, but, instead, remain at the place of oxygen vacancies and can eventually accumulate in cerium oxide nanocrystals.

The following mechanism of decomposition of hypochlorite anions with the help of cerium oxide nanoparticles can be proposed (Fig. 3.22). As shown in numerous

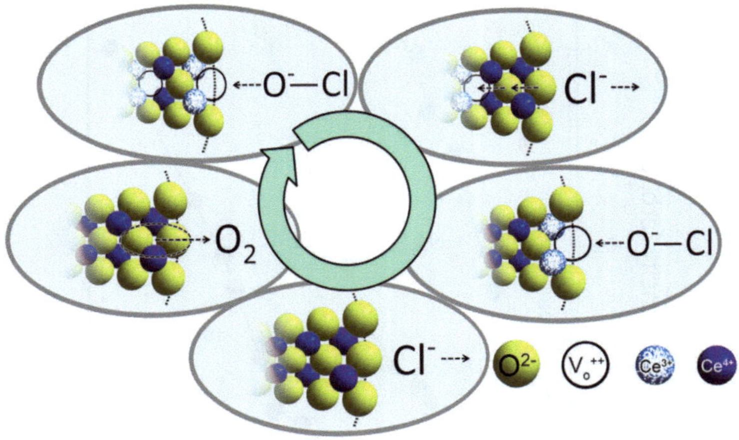

Fig. 3.22 Mechanism of decomposition of hypochlorite anions on the surface of cerium oxide nanocrystals. (Reprinted with permission from *Journal of Physical Chemistry C*, vol. 123, V. Seminko, P. Maksimchuk, G. Grygorova, O. Avrunin, V. Semenets, V. K. Klochkov, Yu.V. Malyukin, Catalytic Decomposition of Hypochlorite Anions by Ceria Nanoparticles Visualized by Spectroscopic Techniques, pp. 20675–20681, 2029. Copyright 2019 American Chemical Society)

publications concerning the catalytic action of cerium oxide, oxygen vacancies on the surface of cerium oxide (primarily on the (111) and (100) surfaces, which are the most catalytically active) can serve as sites for adsorption and subsequent catalytic cleavage different molecules [80, 81]. The interaction of the hypochlorite anion with the catalytic site (Ce^{3+}-Vo-Ce^{3+} complex) on the surface of a cerium oxide nanoparticle leads to the oxidation of adjacent Ce^{3+} ions to Ce^{4+} according to reaction (3.15).

The high degree of oxygen non-stoichiometry of cerium oxide nanoparticles leads to a high mobility of oxygen in these nanocrystals, so that oxygen ions adsorbed by Ce^{3+}-Vo-Ce^{3+} complexes can diffuse into the nanoparticles, leaving the surface free for new acts of splitting hypochlorite anions. Thus, a rapid decrease in the intensity of Ce^{3+} luminescence is observed at the initial stage, since all catalytic sites are free at this stage, while the subsequent rate of catalytic cleavage strongly depends on the processes of removing oxygen from the surface catalytic sites, i.e., it is controlled by the mobility of oxygen in cerium oxide nanoparticles. Diffusion of oxygen in cerium oxide nanoparticles leads to oxidation of deep-seated Ce^{3+} ions, and the higher the hypochlorite concentration, the greater the amount of Ce^{3+} ions undergoing $Ce^{3+} \rightarrow Ce^{4+}$ transitions (up to 87% for 16 mM NaOCl).

The position of the transition point from the predominance of oxygen accumulation processes accompanied by $Ce^{3+} \rightarrow Ce^{4+}$ transitions to the predominance of oxygen release processes accompanied by the reverse $Ce^{4+} \rightarrow Ce^{3+}$ transition is determined by a change in the balance between the concentration of hypochlorite molecules, the number of Ce^{3+}-Vo-Ce^{3+} complexes on the surface of nanoparticles, and the oxygen capacity of nanoparticles. The absorption of oxygen brings the nanoparticle out of thermodynamic equilibrium due to the occupation of oxygen vacancies by oxygen. Thus, when the concentration of hypochlorite decreases, part of the surface catalytic complexes Ce^{3+}-Vo-Ce^{3+} remains inactive and the process of gradual oxygen release begins. Oxygen is released from nanoparticles in a neutral form, so that each oxygen atom leaves behind two electrons, which are captured again by two Ce^{4+} ions. These oxygen atoms form oxygen molecules, increasing the level of dissolved oxygen observed experimentally.

At a certain moment, the concentration of hypochlorite drops to the level where it can no longer affect the overall process, and the subsequent dynamics of the Ce^{3+} luminescence intensity change is determined solely by the release of oxygen from the nanoparticles. The rate of oxygen release at this stage is almost independent of the presence of hypochlorite remaining in the water and is determined only by the mobility of oxygen in cerium oxide nanocrystals. The rate of oxygen diffusion at this stage is significantly lower than at the stage of oxygen accumulation, since the absorption of oxygen by nanoparticles leads to a decrease in the degree of non-stoichiometry of cerium oxide nanoparticles.

So, the general scheme of the catalytic decomposition of hypochlorite by cerium oxide nanoparticles in aqueous colloidal solutions includes the oxidation of two Ce^{3+} ions by hypochlorite anions, which occurs at the Ce^{3+}-Vo-Ce^{3+} catalytic site with subsequent accumulation of oxygen (O^{2-}) by cerium oxide nanoparticles. The balance between the accumulation and release of oxygen is determined both by the thermodynamics of nanoparticles (the concentration of oxygen vacancies) and by the concentration of hypochlorite. A decrease in the concentration of hypochlorite leads to the gradual release of oxygen from nanoparticles, which is accompanied by the recovery of the initial content of Ce^{3+} ions.

3.2.3 *Interaction of Cerium Oxide Nanocrystals with Hydroxyl Radicals (˙OH)*

In Sect. 3.2.1, it was noted that the catalase-like activity of cerium oxide nanocrystals should be accompanied by the ability to neutralize hydroxyl radicals (˙OH), which, along with the interaction of hydrogen peroxide with two Ce^{3+} ions at once on the Ce^{3+}-Vo-Ce^{3+} catalytic sites, is one of the key mechanisms for preventing the generation of significant concentrations of hydroxyl radicals as a result of the Fenton reaction (3.5).

An increase in the content of hydroxyl radicals (˙OH), which are the strongest oxidants among all ROS, can provoke pathological processes inside the cell from increased peroxidation of cell membrane lipids to DNA damage [82]. In general, hydroxyl radicals are formed during the radiolysis of water during X-ray or gamma irradiation ($H_2O \rightarrow H^{\cdot} + {}^{\cdot}OH$) or from hydrogen peroxide in the Fenton reaction ($Fe^{2+} + H_2O_2 \rightarrow Fe^{3+} + {}^{\cdot}OH + OH^-$) and similar reactions catalyzed by metal ions [20]. Since cells mainly consist of water, high-energy radiation leads to the formation of a significant number of hydroxyl radicals in the cell, which, in turn, is one of the key factors determining the danger of high-energy radiation (X-ray or gamma radiation) for the human body. In addition, due to the presence of significant concentrations of hydrogen peroxide inside a living cell (which, as mentioned earlier, plays an important role in the transmission of intracellular signals), the presence of metal ions with variable valence can lead to the formation of hydroxyl radicals by a Fenton-like mechanism. Thus, in the cell, ions with variable valence, which are a key part of a large part of biologically important molecules (in particular, hemoglobin molecules and a number of biological enzymes) are present exclusively in a bound form.

The extremely high redox activity of $^{\cdot}OH$ radicals (the average lifetime of $^{\cdot}OH$ in the biological environment is only a few nanoseconds) complicates its effective elimination by the internal systems of a living cell, which causes the inefficiency of biological enzymes (such as SOD and catalase) in neutralization of hydroxyl radicals inside the cell. Thus, the neutralization of hydroxyl radicals in living cells occurs mainly with the involvement of antioxidant molecules (such as ascorbic acid), which, in turn, is also not particularly effective due to the fact that the concentration of these molecules in cells is limited, and their recovery after interaction with hydroxyl radicals occurs with the participation of intracellular glutathione, the content of which in cells is also limited.

The ability of cerium oxide nanocrystals to protect living cells from the action of hydroxyl radicals allows us to propose these nanocrystals as a new type of antioxidants, which ability to recover their antioxidant properties after interaction with hydroxyl radicals favorably distinguishes them from antioxidant molecules (such as ascorbic acid). Previously, the antiradical effect of cerium oxide nanocrystals (in particular, in relation to hydroxyl radicals) was shown in a number of studies on biological and model systems. In [83], during the study of the oxidation of methylene violet dye with a mixture of Fe^{2+} and H_2O_2 with and without cerium oxide nanoparticles, effective neutralization of hydroxyl radicals by nanoparticles was found. The dependence of the antioxidant effect of cerium oxide nanocrystals on the size of nanoparticles was also shown. Since the decrease in the size of cerium oxide nanoparticles is accompanied by an increase in the content of Ce^{3+} ions, the authors [83] suggested that the antiradical activity of cerium oxide nanocrystals is directly proportional to the amount of Ce^{3+} ions on the surface of the nanoparticle. However, other authors [84] have shown that cerium oxide nanocrystals, in addition to scavenging $^{\cdot}OH$-radicals, can also generate them from hydrogen peroxide by a reaction similar to the Fenton reaction. The opposite results obtained in works [83, 84] indicate that the redox activity of cerium oxide nanoparticles strongly depends on the method of their production, as well as on the properties of the environment (for example, on the pH value of the environment).

The role of surface defects (Ce^{3+}/Ce^{4+} ions, Ce^{3+}-Vo-Ce^{3+}complexes, oxygen vacancies) in the formation of the antiradical activity of nanocrystals was studied for cerium oxide and doped cerium oxide nanocrystals.

As can be seen from Fig. 3.23, both aliovalent impurity ions (Y^{3+}, Eu^{3+}, Tb^{3+}) and isovalent ions of smaller radius (Zr^{4+}) lead to an increase in luminescence intensity

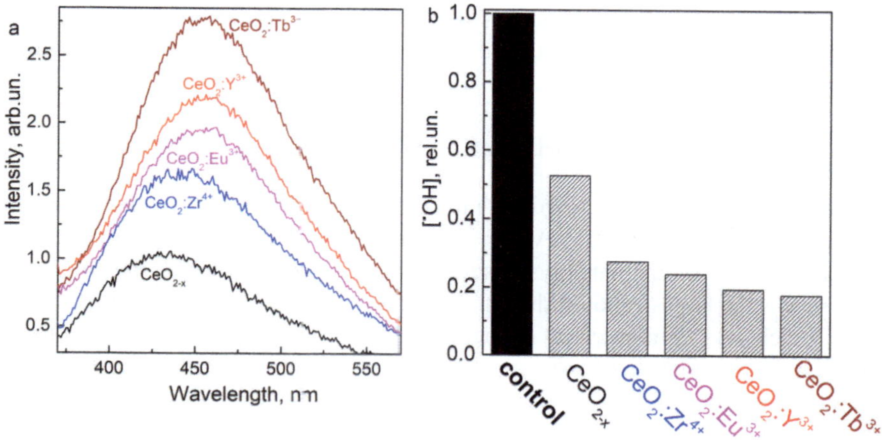

Fig. 3.23 Luminescence spectra of CeO_{2-x} nanocrystals (2 nm) and doped nanocrystals (**a**) and ˙OH scavenging by these nanocrystals (**b**). (Reprinted with permission from *Chemical Physics Letters*, vol. 767, V. Seminko, P. Maksimchuk, G. Grygorova, E. Okrushko, O. Sedyh, O. Avrunin, V. Semenets, Y. Malyukin, Switching the type of redox activity of colloidal nanoceria by Re^{3+} (Re = Y, Eu, Tb) doping, pp. 138363, 2021. Copyright 2021 Elsevier)

of Ce^{3+} ions. The calculations from luminescence spectra show that the content of Ce^{3+} ions in doped 2 nm nanocrystals is 15% for CeO_2, 22% for Zr^{4+}-doped, 25% for Eu^{3+}-doped, 30% for Y^{3+}-doped, and 35% for Tb^{3+}-doped nanocrystals.

As was shown in Sect. 3.2.1., the ability of doped nanocrystals to decompose hydrogen peroxide is determined not by the content of Ce^{3+} ions, but by the content of Ce^{3+}-Vo-Ce^{3+} complexes. Instead, the ability of nanocrystals to neutralize hydroxyl radicals is determined solely by the content of Ce^{3+} ions. Figure 3.23 shows that the presence of cerium oxide nanocrystals or doped oxides in solutions leads to a significant decrease in the content of ˙OH-radicals. So, CeO_2:Tb^{3+} nanocrystals, in which the content of Ce^{3+} ions is the largest (according to the luminescence spectra in Fig. 3.23a), showed the greatest antiradical activity. So, the presence of impurity ions in cerium oxide nanocrystals leads to an increase in the ability to scavenge ˙OH-radicals, which correlates with an increase in the content of Ce^{3+} ions.

In general, the results of the experiments indicate that presence of impurity ions leads to a simultaneous increase in the content of Ce^{3+} ions and to improvement of antiradical properties. An increase in the Ce^{3+} content in the presence of aliovalent Re^{3+} ions (Re = Y, Eu, Tb) is due to formation of Ce^{3+}-Vo-Re^{3+} complexes. The number of such complexes at a low concentration of Re^{3+} ions (10 at.% or less) should be proportional to the concentration of aliovalent ions, while at higher concentrations Re^{3+}-Vo-Re^{3+} complexes are dominant. When Zr^{4+} ions are present, an increase in the concentration of Ce^{3+} is associated with an increase in the content of oxygen vacancies due to a decrease in the energy of oxygen vacancy formation near the Zr^{4+} ion, which in turn leads to the formation of additional Ce^{3+}-Vo-Ce^{3+} complexes.

The mechanism of scavenging of $^{\cdot}OH$-radicals by cerium oxide nanocrystals was proposed in [83] and consists of the oxidation of Ce^{3+} ions by hydroxyl radicals in the following way:

$$^{\cdot}OH + Ce^{3+} \rightarrow OH^{-} + Ce^{4+} \quad (3.16)$$

Thus, unlike hydrogen peroxide molecules or hypochlorite anions, hydroxyl radicals interact not with Ce^{3+}-Vo-Ce^{3+} or Ce^{3+}-Vo-Re^{3+} complexes, but with individual Ce^{3+} ions on the nanocrystal surface. Nanocrystals, in which the content of Ce^{3+} ions is the largest, should show the greatest ability to neutralize hydroxyl radicals, which was demonstrated in the experiment. The impurity ions lead to an increase in the content of oxygen vacancies in cerium oxide nanocrystals, and, therefore, in the amount of Ce^{3+} ions improving the antiradical properties of nanoparticles.

3.2.4 Interaction of Cerium Oxide Nanocrystals with Superoxide Anions (O_2^-)

Superoxide anions are one of the most widespread types of ROS in a living cell. Superoxide anions (O_2^-) are formed in large concentrations in a living cell due to the imperfection of the electron transport chain in the respiratory system of the cell, which leads to the possible capture of electrons by oxygen molecules with the formation of superoxide anions according to the reaction:

$$O_2 + e^{-} \rightarrow O_2^{-} \quad (3.17)$$

Superoxide anions are much less reactive than hydroxyl radicals, but they can easily transform into other ROS, which can be harmful to a living cell [85, 86]. In addition, superoxide anions can enhance the ability of ions with variable valence to generate hydroxyl radicals by the Fenton mechanism due to their role as a reducing agent.

The level of superoxide anions in a living cell is controlled by the superoxide dismutase enzyme (SOD), which converts superoxide anions into hydrogen peroxide without the formation of hydroxyl radicals as intermediate products. According to a number of publications, cerium oxide nanocrystals can also exhibit a SOD-like effect [27, 50].

In order to determine the ability of cerium oxide nanocrystals for SOD-like activity, the autoxidation reaction of adrenaline to adrenochrome was used. This reaction is accompanied by the formation of superoxide anions at the initial stage, which are used for oxidation at the next stage of the reaction. Thus, the addition of any agent that intercepts superoxide anions should lead to inhibition of this reaction, and, therefore, to decrease in the content of the final product of the reaction, i.e., the adrenochrome. Adrenochrome, in turn, is characterized by intense fluorescence

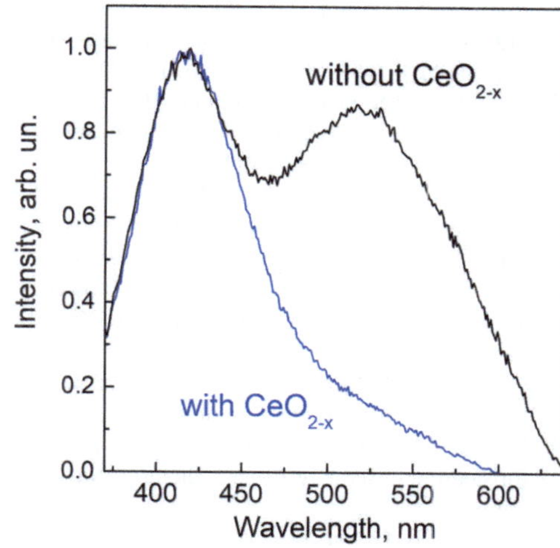

Fig. 3.24 Luminescence spectra of adrenaline without and with cerium oxide nanocrystals (2 nm). The spectra show a decrease in the content of adrenochrome (fluorescent product of autoxidation of adrenaline) when cerium oxide nanocrystals are added

with a maximum at 520 nm, which makes it possible to use optical spectroscopy methods to study the SOD-like activity of nanocrystals.

Figure 3.24 shows that the addition of cerium oxide nanocrystals (2 nm) significantly reduces the fluorescence intensity of adrenochrome (520 nm). This result indicates the effective neutralization of superoxide anions by cerium oxide nanocrystals, which correlates with the results obtained in [27, 37], where it was shown that the biological activity of cerium oxide nanocrystals is similar to the activity of superoxide dismutase.

At the same time, it should be noted that the luminescence of Ce^{3+} ions is much less intense than the fluorescence intensity of adrenaline, which is in the same spectral range, preventing the use of the Ce^{3+} luminescence dynamics to determine the detailed mechanisms of the antioxidant (SOD-like) activity of cerium oxide nanocrystals. An alternative approach to this problem will be proposed in the next chapter.

References

1. Rzigalinski BA (2005) Nanoparticles and cell longevity. Technol Cancer Res Treat 4:651–659. https://doi.org/10.1177/153303460500400609
2. Tarnuzzer RW, Colon J, Patil S, Seal S (2005) Vacancy engineered ceria nanostructures for protection from radiation-induced cellular damage. Nano Lett 5:2573–2577. https://doi.org/10.1021/nl052024f
3. Das M, Patil S, Bhargava N, Kang JF, Riedel LM, Seal S, Hickman JJ (2007) Auto-catalytic ceria nanoparticles offer neuroprotection to adult rat spinal cord neurons. Biomaterials 28:1918–1925. https://doi.org/10.1016/j.biomaterials.2006.11.036
4. Perez JM, Asati A, Nath S, Kaittanis C (2008) Synthesis of biocompatible dextran-coated nanoceria with pH-dependent antioxidant properties. Small 4:552–556. https://doi.org/10.1002/smll.200700824

5. Karakoti AS, Singh S, Kumar A, Malinska M, Kuchibhatla SV, Wozniak K, Seal S (2009) PEGylated nanoceria as radical scavenger with tunable redox chemistry. J Am Chem Soc 131:14144–14145. https://doi.org/10.1021/ja9051087
6. Chen J, Patil S, Seal S, McGinnis JF (2008) Nanoceria particles prevent ROI-induced blindness. In: Recent advances in retinal degeneration. Springer, New York, pp 53–59. https://doi.org/10.1007/978-0-387-74904-4_5
7. Kong L, Cai X, Zhou X, Wong LL, Karakoti AS, Seal S, McGinnis JF (2011) Nanoceria extend photoreceptor cell lifespan in tubby mice by modulation of apoptosis/survival signaling pathways. Neurobiol Dis 42:514–523. https://doi.org/10.1016/j.nbd.2011.03.004
8. Das S, Chigurupati S, Dowding J, Munusamy P, Baer DR, McGinnis JF, Seal S (2014) Therapeutic potential of nanoceria in regenerative medicine. MRS Bull 39:976–983. https://doi.org/10.1557/mrs.2014.221
9. Chen J, Seal S, Sezate SA, Ramsey M, Elias R, Wong L, McGinnis JF (2005) Nanoceria particles confer neuroprotection in retinal cells in vitro. Invest Ophthalmol Vis Sci 43:186–186
10. Estevez AY, Erlichman JS (2014) The potential of cerium oxide nanoparticles (nanoceria) for neurodegenerative disease therapy. Nanomedicine 9:1437–1440. https://doi.org/10.2217/nnm.14.87
11. Halliwell B, Gutteridge JM (2015) Free radicals in biology and medicine. Oxford University Press, Oxford
12. Apel K, Hirt H (2004) Reactive oxygen species: metabolism, oxidative stress, and signal transduction. Annu Rev Plant Biol 55:373–399. https://doi.org/10.1146/annurev.arplant.55.031903.141701
13. Lee J, Koo N, Min DB (2004) Reactive oxygen species, aging, and Antioxidative nutraceuticals. Compr Rev Food Sci Food Saf 3:21–33. https://doi.org/10.1111/j.1541-4337.2004.tb00058.x
14. Papa S, Skulachev VP (1997) Reactive oxygen species, mitochondria, apoptosis and aging. In: Detection of mitochondrial diseases. Springer, Boston, MA, pp 305–319. https://doi.org/10.1007/978-1-4615-6111-8_47
15. Stadtman ER, Berlett BS (1997) Reactive oxygen-mediated protein oxidation in aging and disease. Chem Res Toxicol 10:485–494. https://doi.org/10.1021/tx960133r
16. Rhee SG (2006) H_2O_2, a necessary evil for cell signaling. Science 312:1882–1883. https://doi.org/10.1126/science.1130481
17. Sies H (2014) Role of metabolic H_2O_2 generation in redox signaling and oxidative stress. J Biol Chem 289:8735–8741. https://doi.org/10.1074/jbc.R113.544635
18. Cadet J, Delatour T, Douki T, Gasparutto D, Pouget JP, Ravanat JL, Sauvaigo S (1999) Hydroxyl radicals and DNA Base damage. Mutat Res Fundam Mol Mech Mutagen 424:9–21. https://doi.org/10.1016/s0027-5107(99)00004-4
19. Poeggeler B, Reiter RJ, Tan DX, Chen LD, Manchester LC (1993) Melatonin, hydroxyl radical-mediated oxidative damage, and aging: a hypothesis. J Pineal Res 14:151–168. https://doi.org/10.1111/j.1600-079X.1993.tb00498.x
20. Wardman P, Candeias LP (1996) Fenton chemistry: an introduction. Radiat Res 145:523–531
21. Halliwell B, Gutteridge J (1984) Oxygen toxicity, oxygen radicals, transition metals and disease. Biochem J 219:1. https://doi.org/10.1042/bj2190001
22. Chevion M (1988) A site-specific mechanism for free radical induced biological damage: the essential role of redox-active transition metals. Free Radic Biol Med 5:27–37. https://doi.org/10.1016/0891-5849(88)90059-7
23. Vidrio E, Jung H, Anastasio C (2008) Generation of hydroxyl radicals from dissolved transition metals in surrogate lung fluid solutions. Atmos Environ 42:4369–4379. https://doi.org/10.1016/j.atmosenv.2008.01.004
24. Matés JM, Pérez-Gómez C, De Castro IN (1999) Antioxidant enzymes and human diseases. Clin Biochem 32:595–603. https://doi.org/10.1016/S0009-9120(99)00075-2
25. Sun Y (1990) Free radicals, antioxidant enzymes, and carcinogenesis. Free Radic Biol Med 8:583–599. https://doi.org/10.1016/0891-5849(90)90156-D

26. Gutteridge JM, Halliwell B (2010) Antioxidants: molecules, medicines, and myths. Biochem Biophys Res Commun 393:561–564. https://doi.org/10.1016/j.bbrc.2010.02.071
27. Heckert EG, Karakoti AS, Seal S, Self WT (2008) The role of cerium redox state in the SOD mimetic activity of nanoceria. Biomaterials 29:2705–2709. https://doi.org/10.1016/j.biomaterials.2008.03.014
28. Pirmohamed T, Dowding JM, Singh S, Wasserman B, Heckert EG, Karakoti AS, Self WT (2010) Nanoceria exhibit redox state-dependent catalase mimetic activity. Chem Commun 46:2736–2738. https://doi.org/10.1039/3922024K
29. Wei H, Wang E (2013) Nanomaterials with enzyme-like characteristics (nanozymes): next-generation artificial enzymes. Chem Soc Rev 42:6060–6093. https://doi.org/10.1039/C3CS35486E
30. Wang X, Hu Y, Wei H (2016) Nanozymes in bionanotechnology: from sensing to therapeutics and beyond. Inorg Chem Front 3:41–60. https://doi.org/10.1039/C5QI00240K
31. Huang Y, Liu Z, Liu C, Ju E, Zhang Y, Ren J, Qu X (2016) Self-assembly of multi-nanozymes to mimic an intracellular antioxidant defense system. Angew Chem Int Ed 55:6646–6650. https://doi.org/10.1002/anie.201600868
32. Singh N, Savanur MA, Srivastava S, D'Silva P, Mugesh G (2017) A redox modulatory Mn_3O_4 nanozyme with multi-enzyme activity provides efficient cytoprotection to human cells in a Parkinson's disease model. Angew Chem Int Ed 56:14267–14271. https://doi.org/10.1002/anie.201708573
33. Wu J, Wang X, Wang Q, Lou Z, Li S, Zhu Y, Wei H (2019) Nanomaterials with enzyme-like characteristics (nanozymes): next-generation artificial enzymes (II). Chem Soc Rev 48:1004–1076. https://doi.org/10.1039/C8CS00457A
34. Cheng H, Lin S, Muhammad F, Lin YW, Wei H (2016) Rationally modulate the oxidase-like activity of nanoceria for self-regulated bioassays. ACS Sens 1:1336–1343. https://doi.org/10.1021/acssensors.6b00500
35. Guo W, Zhang M, Lou Z, Zhou M, Wang P, Wei H (2019) Engineering nanoceria for enhanced peroxidase mimics: a solid solution strategy. ChemCatChem 11:737–743. https://doi.org/10.1002/cctc.201801578
36. Zholobak NM, Ivanov VK, Shcherbakov AB, Shaporev AS, Polezhaeva OS, Baranchikov AY, Tretyakov YD (2011) UV-shielding property, photocatalytic activity and photocytotoxicity of ceria colloid solutions. J Photochem Photobiol B 102:32–38. https://doi.org/10.1016/j.jphotobiol.2010.09.002
37. Singh S, Dosani T, Karakoti AS, Kumar A, Seal S, Self WT (2011) A phosphate-dependent shift in redox state of cerium oxide nanoparticles and its effects on catalytic properties. Biomaterials 32:6745–6753. https://doi.org/10.1016/j.biomaterials.2011.05.073
38. Estevez AY, Pritchard S, Harper K, Aston JW (2011) Neuroprotective mechanisms of cerium oxide nanoparticles in a mouse hippocampal brain slice model of ischemia. Free Radic Biol Med 51:1155–1163. https://doi.org/10.1016/j.freeradbiomed.2011.06.006
39. Lee SS, Song W, Cho M, Puppala HL, Nguyen P, Zhu H, Colvin VL (2013) Antioxidant properties of cerium oxide nanocrystals as a function of nanocrystal diameter and surface coating. ACS Nano 7:9693–9703. https://doi.org/10.1021/nn4026806
40. Kim CK, Kim T, Choi IY, Soh M, Kim D, Kim YJ, Park SP (2012) Ceria nanoparticles that can protect against ischemic stroke. Angew Chem Int Ed 51:11039–11043. https://doi.org/10.1002/anie.201203780
41. Tsai YY, Oca-Cossio J, Lin SM, Woan K, Yu PC, Sigmund W (2008) Reactive oxygen species scavenging properties of ZrO_2–CeO_2 solid solution nanoparticles. Nanomedicine 3:637–645. https://doi.org/10.2217/17435889.3.5.637
42. Shcherbakov AB, Zholobak NM, Ivanov VK (2020) Biological, biomedical and pharmaceutical applications of cerium oxide. In: cerium oxide (CeO_2): synthesis, properties and applications. Elsevier, pp 279–358. https://doi.org/10.1016/B978-0-12-815661-2.00008-6
43. Celardo I, Pedersen JZ, Traversa E, Ghibelli L (2011) Pharmacological potential of cerium oxide nanoparticles. Nanoscale 3:1411–1420. https://doi.org/10.1039/C0NR00875C

44. Popov AL, Shcherbakov AB, Zholobak NM, Baranchikov AY, Ivanov VK (2017) Cerium dioxide nanoparticles as third-generation enzymes (nanozymes). Nanosyst Phys Chem Math 8:760–781. https://doi.org/10.17586/2220-8054-2017-8-6-760-781
45. Shcherbakov AB, Zholobak NM, Baranchikov AE, Ryabova AV, Ivanov VK (2015) Cerium fluoride nanoparticles protect cells against oxidative stress. Mater Sci Eng C 50:151–159. https://doi.org/10.1016/j.msec.2015.01.094
46. Plakhova TV, Romanchuk AY, Butorin SM, Konyukhova AD, Egorov AV, Shiryaev AA, Bauters S (2019) Towards the surface hydroxyl species in CeO_2 nanoparticles. Nanoscale 11:18142–18149. https://doi.org/10.1039/C9NR06032D
47. Cafun JD, Kvashnina KO, Casals E, Puntes VF, Glatzel P (2013) Absence of Ce^{3+} sites in chemically active colloidal ceria nanoparticles. ACS Nano 7:10726–10732. https://doi.org/10.1021/nn403542p
48. Plakhova TV, Romanchuk AY, Yakunin SN, Dumas T, Demir S, Wang S (2016) Solubility of nanocrystalline cerium dioxide: experimental data and thermodynamic modeling. J Phys Chem C 120:22615–22626. https://doi.org/10.1021/acs.jpcc.6b05650
49. Celardo I, De Nicola M, Mandoli C, Pedersen JZ, Traversa E, Ghibelli L (2011) Ce^{3+} ions determine redox-dependent anti-apoptotic effect of cerium oxide nanoparticles. ACS Nano 5:4537–4549. https://doi.org/10.1021/nn200126a
50. Karakoti A, Singh S, Dowding JM, Seal S, Self WT (2010) Redox-active radical scavenging nanomaterials. Chem Soc Rev 39:4422–4432. https://doi.org/10.1039/B919677N
51. Yokel RA, Hussain S, Garantziotis S, Demokritou P, Castranova V, Cassee FR (2014) The yin: an adverse health perspective of nanoceria: uptake, distribution, accumulation, and mechanisms of its toxicity. Environ Sci Nano 1:406–428. https://doi.org/10.1039/C4EN00039K
52. Kumar A, Das S, Munusamy P, Self W, Baer DR, Sayle DC, Seal S (2014) Behavior of nanoceria in biologically relevant environments. Environ Sci Nano 1:516–532. https://doi.org/10.1039/C4EN00052H
53. Andreescu D, Bulbul G, Özel RE, Hayat A, Sardesai N, Andreescu S (2014) Applications and implications of nanoceria reactivity: measurement tools and environmental impact. Environ Sci Nano 1:445–458. https://doi.org/10.1039/C4EN00075G
54. Dhall A, Self W (2018) Cerium oxide nanoparticles: a brief review of their synthesis methods and biomedical applications. Antioxidants 7:97. https://doi.org/10.3390/antiox7080097
55. Li C, Shi X, Shen Q, Guo C, Hou Z, Zhang J (2018) Hot topics and challenges of regenerative nanoceria in application of antioxidant therapy. J Nanomater 2018:4857461. https://doi.org/10.1155/2018/4857461
56. Dvořák F, Szabová L, Johánek V, Farnesi Camellone M, Stetsovych V, Vorokhta M, Mysliveček J (2018) Bulk hydroxylation and effective water splitting by highly reduced cerium oxide: the role of O vacancy coordination. ACS Catal 8:4354–4363. https://doi.org/10.1021/acscatal.7b04409
57. Venstrom LJ, Petkovich N, Rudisill S, Stein A, Davidson JH (2012) The effects of morphology on the oxidation of ceria by water and carbon dioxide. J Sol Energy Eng 134:014505. https://doi.org/10.1115/1.4005119
58. Parent AR, Crabtree RH, Brudvig GW (2013) Comparison of primary oxidants for water-oxidation catalysis. Chem Soc Rev 42:2247–2252. https://doi.org/10.1039/C2CS35225G
59. Kuramoto Y (2003) Chemical oscillations, waves, and turbulence. Courier Corp.
60. Nicolis G, Portnow J (1973) Chemical oscillations. Chem Rev 73:365–384. https://doi.org/10.1021/cr60284a003
61. Zhabotinsky AM (1991) A history of chemical oscillations and waves. Chaos 1:379–386. https://doi.org/10.1063/1.165848
62. Urey HC, Dawsey LH, Rice FO (1929) The absorption spectrum and decomposition of hydrogen peroxide by light. J Am Chem Soc 51:1371–1383. https://doi.org/10.1021/ja01380a011
63. Anbar M, Dostrovsky I (1954) Ultra-violet absorption spectra of some organic hypohalites. J Chem Soc:1105–1108. https://doi.org/10.1039/JR9540001105

64. Payá M, Halliwell B, Hoult JRS (1992) Interactions of a series of coumarins with reactive oxygen species: scavenging of superoxide, hypochlorous acid and hydroxyl radicals. Biochem Pharmacol 44:205–214. https://doi.org/10.1016/0006-2952(92)90002-z
65. Sun M, Zigman S (1978) An improved spectrophotometric assay for superoxide dismutase based on epinephrine autoxidation. Anal Biochem 90:81–89. https://doi.org/10.1016/0003-2697(78)90010-6
66. Grulke E, Reed K, Beck M, Huang X, Cormack A, Seal S (2014) Nanoceria: factors affecting its pro-and anti-oxidant properties. Environ Sci Nano 1:429–444. https://doi.org/10.1039/C4EN00105B
67. Michaelis L, Menten ML (1913) Die Kinetik der Invertinwirkung. Biochem Z 49:333–369
68. Yu P, Hayes SA, O'Keefe TJ, O'Keefe MJ, Stoffer JO (2006) The phase stability of cerium species in aqueous systems II. The systems. Equilibrium considerations and Pourbaix diagram calculations. J Electrochem Soc 153:C74–C79. https://doi.org/10.1149/1.2130572
69. Pullar JM, Vissers MC, Winterbourn CC (2000) Living with a killer: the effects of hypochlorous acid on mammalian cells. IUBMB Life 50:259–266. https://doi.org/10.1080/713803731
70. Babior BM (1984) The respiratory burst of phagocytes. J Clin Invest 73:599–601. https://doi.org/10.1172/JCI111249
71. Dahlgren C, Karlsson A (1999) Respiratory burst in human neutrophils. J Immunol Methods 232:3–14. https://doi.org/10.1016/S0022-1759(99)00146-5
72. Baggiolini M, Wymann MP (1990) Turning on the respiratory burst. Trends Biochem Sci 15:69–72. https://doi.org/10.1016/0968-0004(90)90179-f
73. Schraufstätter IU, Browne K, Harris A, Hyslop PA, Jackson JH, Quehenberger O, Cochrane C (1990) Mechanisms of hypochlorite injury of target cells. J Clin Invest 85:554–562. https://doi.org/10.1172/jci114472
74. Spencer JP, Whiteman M, Jenner A, Halliwell B (2000) Nitrite-induced deamination and hypochlorite induced oxidation of DNA in intact human respiratory tract epithelial cells. Free Radic Biol Med 28:1039–1050. https://doi.org/10.1016/S0891-5849(00)00190-8
75. Carr AC, van den Berg JJM, Winterbourn CC (1996) Chlorination of cholesterol in cell membranes by hypochlorous acid. Arch Biochem Biophys 332:63–69. https://doi.org/10.1006/abbi.1996.0317
76. Armstrong DA, Buchanan JD (1978) Reactions of O_2^-, H_2O_2 and other oxidants with sulfhydryl enzymes. Photochem Photobiol 28:743–755. https://doi.org/10.1111/j.1751-1097.1978.tb07011.x
77. Grisham MB, Jefferson MM, Melton DF, Thomas EL (1984) chlorination of endogenous amines by isolated neutrophils. Ammonia-dependent bactericidal, cytotoxic and cytolytic activities of the chloramines. J Biol Chem 259:10404–10413. https://doi.org/10.1016/S0021-9258(18)90979-8
78. Malech HL, Gallin JI (1987) Neutrophils in human diseases. N Engl J Med 317:687–694. https://doi.org/10.1056/NEJM198709103171107
79. Pulido-Reyes G, Das S, Leganés F, Silva SO, Wu S, Self W, Seal S (2016) Hypochlorite scavenging activity of cerium oxide nanoparticles. RSC Adv 6:62911–62915. https://doi.org/10.1039/C6RA11131A
80. Jampaiah D, Venkataswamy P, Coyle VE, Reddy BM, Bhargava SK (2016) Low-temperature CO oxidation over manganese, cobalt, and nickel doped CeO_2 nanorods. RSC Adv 6:80541–80548. https://doi.org/10.1039/C6RA13577C
81. Wada S, Oka K, Watanabe K, Izumi Y (2013) Catalytic conversion of carbon dioxide into dimethyl carbonate using reduced copper-cerium oxide catalysts as low as 353 K and 1.3 MPa and the reaction mechanism. Front Chem 1:8. https://doi.org/10.3389/fchem.2013.00008
82. Cadet J, Delatour T, Douki T, Gasparutto D, Pouget JP, Ravanat JL, Sauvaigo S (1999) Hydroxyl radicals and DNA base damage. Mutat Res Fundam Mol Mech Mutagen 424:9–21. https://doi.org/10.1016/S0027-5107(99)00004-4
83. Xue Y, Luan Q, Yang D, Yao X, Zhou K (2011) Direct evidence for hydroxyl radical scavenging activity of cerium oxide nanoparticles. J Phys Chem C 115:4433–4438. https://doi.org/10.1021/jp109819u

84. Lu M, Zhang Y, Wang Y, Jiang M, Yao X (2016) Insight into several factors that affect the conversion between antioxidant and oxidant activities of nanoceria. ACS Appl Mater Interfaces 8:23580–23590. https://doi.org/10.1021/acsami.6b08219
85. Fridovich I (1997) Superoxide anion radical (O_2^-), superoxide dismutases, and related matters. J Biol Chem 272:18515–18517. https://doi.org/10.1074/jbc.272.30.18515
86. Fridovich I (1986) Biological effects of the superoxide radical. Arch Biochem Biophys 247:1–11. https://doi.org/10.1016/0003-9861(86)90526-6

Chapter 4
Control of the Defect Structure and Redox Properties of CeO_{2-x} Nanocrystals

Abstract This chapter explores strategies to control the defect structure and redox properties of CeO_{2-x} nanocrystals, focusing on their role in modulating interaction with reactive oxygen species (ROS). Ultraviolet (UV) irradiation was shown to induce $Ce^{4+} \rightarrow Ce^{3+}$ transitions and increase the content of Ce^{3+}-V_o-Ce^{3+} complexes enhancing the antioxidant activity of nanocrystals against hydroxyl radicals and superoxide anions. The redox behavior of CeO_{2-x} was further demonstrated to be strongly pH-dependent: at neutral and alkaline conditions, catalase- and superoxide dismutase (SOD)-like activities dominate, while in acidic media, prooxidant effects emerge due to hydroxyl radical formation. This dual activity highlights the potential of nanoceria as a selective therapeutic agent, capable of protecting normal cells while promoting oxidative stress in cancer cells. Finally, the chapter discusses the application of CeO_{2-x} and Eu^{3+}-doped CeO_{2-x} nanocrystals as reversible luminescent sensors for hydrogen peroxide, emphasizing the influence of particle size, temperature, and UV irradiation on sensitivity and recovery rates. Together, these findings provide fundamental insights into the tunable redox chemistry of cerium oxide nanocrystals with implications for biomedical, catalytic, and sensing applications.

4.1 The Effect of UV Irradiation on the Defect Structure and Redox Properties of CeO_{2-x} Nanocrystals

4.1.1 Effect of Selective UV Irradiation on the Defect Structure of CeO_{2-x} Nanocrystals

As was shown in Chap. 3, the effectiveness of CeO_{2-x} nanocrystals and mixed oxide nanocrystals as antioxidants directly depends on their defect structure, namely on the content of Ce^{3+}-Vo-Ce^{3+} complexes or Ce^{3+} ions. Since the processes of interaction of reactive oxygen species with the surface of CeO_{2-x} nanocrystals occur precisely on the Ce^{3+}-Vo-Ce^{3+} catalytic sites, an increase in the content of Ce^{3+}-Vo-Ce^{3+} complexes should increase the antioxidant activity of nanoparticles.

V. Seminko, P. Maksimchuk, *Cerium Oxide Nanocrystals (CeO_{2-x})*,
https://doi.org/10.1007/978-3-032-14130-9_4

One of the ways of influencing the defect structure of cerium oxide nanocrystals is long-term irradiation with radiation with wavelengths corresponding to charge-transfer transitions from the valence band to empty $4f^0$ levels of tetravalent cerium ions ($O^{2-} \rightarrow Ce^{4+}$) (Fig. 4.1).

As is known for semiconductors, irradiation with energies more than the band gap value leads to the formation of electron-hole pairs (an electron in the conduction band and a hole in the valence band), which can then either recombine with photon emission or become localized near defects in the crystal lattice, or migrate to the surface of the nanocrystal, where they can interact with various molecules, water or oxygen. The latter is widely used for the needs of photocatalysis, and as it was shown in a number of works, cerium oxide can also act as an effective photocatalytic material [1–5].

It was shown that cerium oxide as a material for photocatalysis can be even better than the most widely used material in this field, titanium oxide (TiO_2), because the lifetime of an electron and a hole in cerium oxide is much higher than in titanium oxide [3]. The instantaneous recombination of an electron and a hole in cerium oxide is complicated by the fact that the transfer of an electron from the valence band to the empty $4f^0$ levels of Ce^{4+} ions is accompanied by a strong localization of the electron. In the majority of works devoted to the processes of electron transport in cerium oxide [6, 7], complexes ($Ce^{4+} + e^-$) are considered as polarons of small radius, the transport of which is carried out exclusively by the hopping mechanism. The electronic structure of the complexes ($Ce^{4+} + e^-$) is generally similar to the electronic structure of Ce^{3+} ions (or completely similar, if we consider the model of electron localization at the levels of cerium ions during the formation of oxygen vacancies), and it can be expected that at a high density of excitation photons, the transition of individual ions from Ce^{4+} to Ce^{3+} as a result of charge transfer should be accompanied by the formation of oxygen vacancies and Ce^{3+}-Vo-Ce^{3+} complexes,

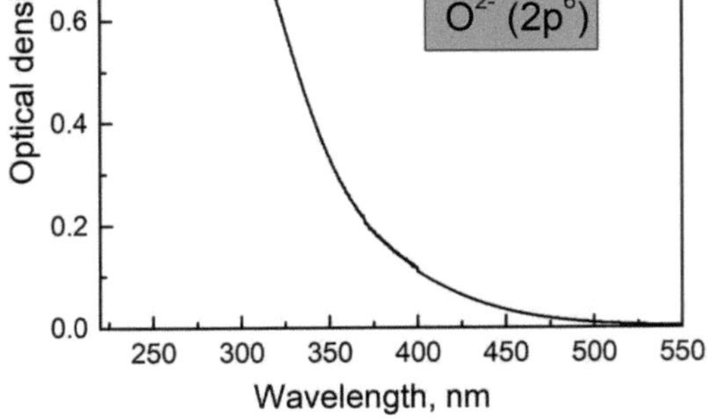

Fig. 4.1 Absorption spectrum of 2 nm cerium oxide nanocrystal and a simplified scheme of energy $O^{2-} \rightarrow Ce^{4+}$ charge transfer

since, as was shown earlier, this configuration is the most stable in the cerium oxide structure.

In order to investigate the effect of UV radiation on the defect structure and redox activity of CeO_{2-x} nanocrystals, broadband UV radiation and laser sources with different wavelengths (325 nm and 530 nm) were used. At broadband UV irradiation by a mercury lamp, a gradual increase in the intensity of Ce^{3+} luminescence was observed (Fig. 4.2), as well as a gradual decrease in the intensity of the $O^{2-} \rightarrow Ce^{4+}$ charge-transfer band on the absorption spectra of aqueous colloidal solutions of cerium oxide nanoparticles (Fig. 4.3). This spectral change was reversible, and after the irradiation, the reverse $Ce^{3+} \rightarrow Ce^{4+}$ process took place over time, which led to decrease in the intensity of the Ce^{3+} luminescence band and an increase in the $O^{2-} \rightarrow Ce^{4+}$ charge-transfer band in the absorption spectra to the values that occurred before irradiation.

Thus, irradiation of CeO_{2-x} nanocrystals with a wavelength corresponding to transitions with charge transfer ($O^{2-} \rightarrow Ce^{4+}$) leads to the capture of electrons by Ce^{4+} ions followed by $Ce^{4+} \rightarrow Ce^{3+}$ transition. At the same time, the fact that after irradiation, there is no fast reverse transition in the form of charge-transfer luminescence or non-radiative relaxation indicates that the obtained complexes are stable confirming that irradiation is accompanied by the formation of Ce^{3+}-Vo-Ce^{3+} complexes providing the stabilization of defects obtained during irradiation.

The effect of laser irradiation on the formation of Ce^{3+}-Vo-Ce^{3+} complexes in the cerium oxide structure was previously shown in [8]. The authors [8] noted that irradiation of cerium oxide powders and films in a vacuum chamber led to a noticeable increase in the luminescence intensity, which was attributed to excitons captured by Ce^{3+}-Vo-Ce^{3+} complexes. The presence of oxygen in the vacuum chamber led to the reverse process. Unexpectedly, for cerium oxide nanocrystals, the effect was the opposite, that is, a decrease in the intensity of luminescence upon irradiation in a

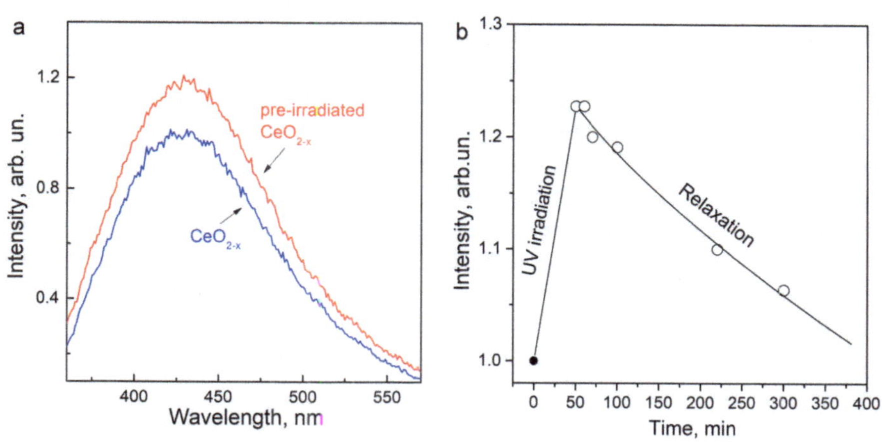

Fig. 4.2 (**a**) Reversible change of the Ce^{3+} luminescence band of colloidal solutions of CeO_{2-x} nanoparticles (2 nm) under UV irradiation. (**b**) Dynamics of changes in the intensity of the Ce^{3+} luminescence band ($\lambda_{reg} = 430$ nm) during irradiation/relaxation (irradiation source–mercury lamp)

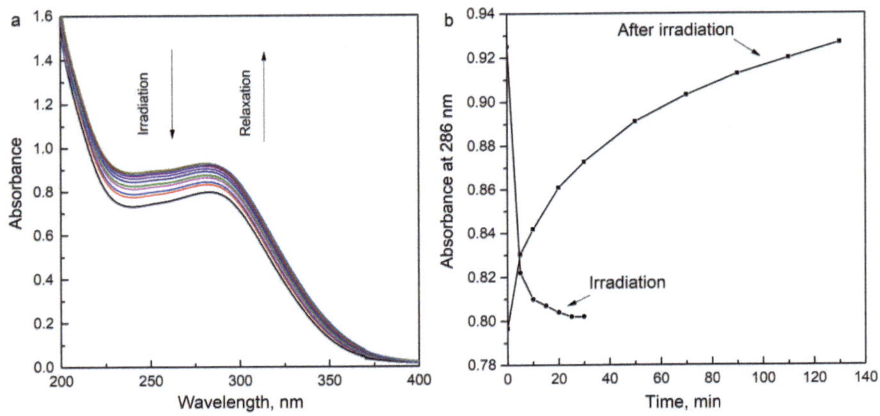

Fig. 4.3 (**a**) Reversible change of charge-transfer band ($O^{2-} \rightarrow Ce^{4+}$) in the absorption spectra of colloidal solutions of CeO_2 nanoparticles (2 nm) at broadband UV irradiation. (**b**) Dynamics of the charge-transfer band intensity (λ_{reg} = 286 nm) during irradiation/relaxation. (Reprinted with permission from *Journal of Photochemistry and Photobiology A: Chemistry*, vol. 364, V.K. Klochkov, Yu.V. Malyukin, G.V. Grygorova, O.O. Sedyh, N.S. Kavok, V.V. Seminko, and V.P. Semynozhenko, Oxidation-reduction processes in CeO_{2-x} nanocrystals under UV irradiation, pp. 282–287, 2018. Copyright 2018 Elsevier)

vacuum chamber with subsequent recovery of luminescence in an oxygen atmosphere. For all samples, photoreduction/photooxidation was observed only at 325 nm laser irradiation, while irradiation with other wavelengths did not change the spectra. The authors [8] proposed two possible mechanisms of the influence of laser irradiation on luminescence processes: either high concentrations and mobility of holes formed during irradiation lead to the possibility of a random meeting of two holes near the O^{2-} ion with subsequent desorption of oxygen and the formation of oxygen vacancies, or 325 nm laser irradiation can lead to the formation of oxygen vacancies on the (100) surfaces of cerium oxide.

A similar effect was observed for bulk $SrTiO_3$ crystals [9] and MgO nanocrystals [10]. Thus, it can be asserted with a high degree of probability that in our experiments, the stability of complexes ($Ce^{4+} + e^-$) obtained upon irradiation with a wavelength corresponding to charge-transfer transitions ($O^{2-} \rightarrow Ce^{4+}$) is explained by strong localization of electrons captured at the $4f^0$ level, as well as by the fact that the process of electron capture can be accompanied by the release of oxygen from the nanoparticle with the formation of Ce^{3+}-Vo-Ce^{3+} complexes. In turn, the content of Ce^{3+}-Vo-Ce^{3+} complexes in cerium oxide nanoparticles should affect the antioxidant activity of nanoparticles when interacting with reactive oxygen species that was verified in a series of experiments described in the next section.

4.1.2 Effect of Selective UV Irradiation on the Antioxidant Properties of CeO_{2-x} Nanocrystals

Since UV irradiation leads to the formation of additional Ce^{3+}-Vo-Ce^{3+} complexes due to the transfer of an electron ($O^{2-} \rightarrow Ce^{4+}$) from an oxygen ion to a cerium ion, this should lead to an increase in the antioxidant activity of cerium oxide nanoparticles, which depends either on the content of Ce^{3+}-Vo-Ce^{3+} complexes (when interacting with hydrogen peroxide or hypochlorite anions) or on the Ce^{3+}/Ce^{4+} ratio (when interacting with hydroxyl radicals or superoxide anions). For all mentioned ROS types, an increase in the content of Ce^{3+}-Vo-Ce^{3+} complexes on the surface of cerium oxide nanoparticles should lead to an increase in the rate of their decomposition in aqueous solutions.

To determine the effect of pre-irradiation on the ability of cerium oxide nanocrystals to scavenge ˙OH radicals, the coumarin test was used. Coumarin has no intrinsic fluorescence, but it is converted to fluorescent 7-hydroxycoumarin during reaction with hydroxyl radicals. Hydroxyl radicals were formed under X-ray irradiation ($U = 30$ kV, $I = 20$ mA), and fluorescence of 7-hydroxycoumarin was used to monitor the ability of nanoparticles to scavenge ˙OH radicals. As was shown in Chap. 3, the ability of cerium oxide nanoparticles to scavenge ˙OH radicals correlates with the content of Ce^{3+} ions on the surface of nanocrystals.

In the experiment, aqueous solutions of cerium oxide nanocrystals were irradiated with 325 nm laser radiation for 1.5 h, and their luminescence and antioxidant properties were compared with non-irradiated samples. In Fig. 4.4a, the luminescence spectra of irradiated and non-irradiated nanocrystals are shown. Laser irradiation leads to a 1.2-fold increase in the intensity of Ce^{3+} luminescence, and the coumarin test shows a 1.4-fold increase in the ability to scavenge ˙OH-radicals in

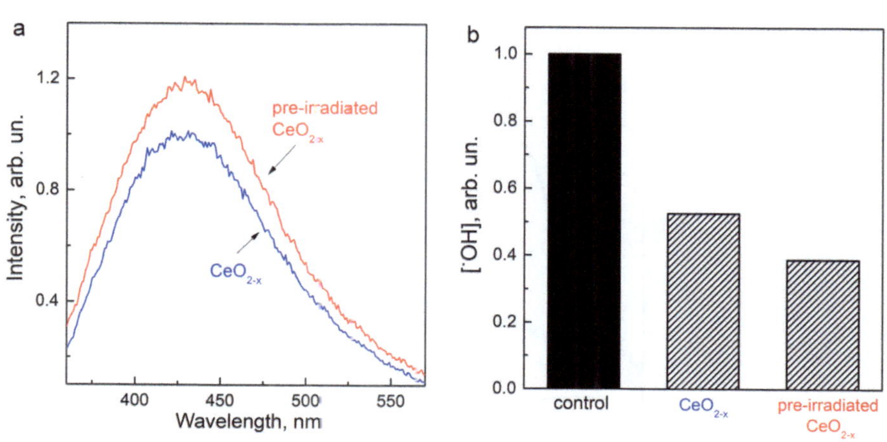

Fig. 4.4 (**a**) Luminescence spectra of CeO_{2-x} nanocrystals (2 nm) without pre-irradiation and pre-irradiated (1.5 h) at $\lambda_{exc} = 325$ nm; (**b**) concentration of ˙OH radicals in aqueous solutions of nanocrystals irradiated with X-ray radiation. Control sample—aqueous solutions without cerium oxide nanocrystals

irradiated samples compared to non-irradiated ones. So, one can conclude that preliminary laser irradiation of colloidal solutions before interaction with ˙OH radicals improves the antiradical properties of nanocrystals due to an increase in the Ce^{3+}/Ce^{4+} ratio at laser irradiation of nanoparticles.

Superoxide anions (O_2^-) are an unstable form of ROS, so the content of superoxide anions in an aqueous solution can be determined only indirectly by oxidation of various organic molecules. Thus, autooxidation of adrenaline is accompanied by the formation of adrenochrome. The autooxidation process is accompanied by the formation of superoxide anions (O_2^-) at the initial stage, which participate afterwards in the autooxidation process. Thus, a decrease in the content of superoxide anions (O_2^-) due to their neutralization by CeO_2 nanocrystals is expected to decrease the content of adrenochrome.

In the experiment, the influence of non-irradiated and previously irradiated aqueous solutions of CeO_{2-x} nanocrystals on the processes of autoxidation of adrenaline was investigated. In Fig. 4.5, the dynamics of growth of adrenochrome (an oxidized product of adrenaline) content for non-irradiated and pre-irradiated aqueous solutions of CeO_{2-x} nanocrystals is shown. It can be seen that pre-irradiation of nanoparticles (1.5 h, 325 nm) increases their ability to neutralize superoxide anions. Thus, it can be concluded that the ability of CeO_{2-x} nanocrystals to neutralize superoxide anions correlates with Ce^{3+}/Ce^{4+} ratio, because as in the case of interaction with hydroxyl radicals, an increase in the Ce^{3+}/Ce^{4+} ratio due to irradiation leads to an increase in the ability of nanocrystals to reduce the content of superoxide anions

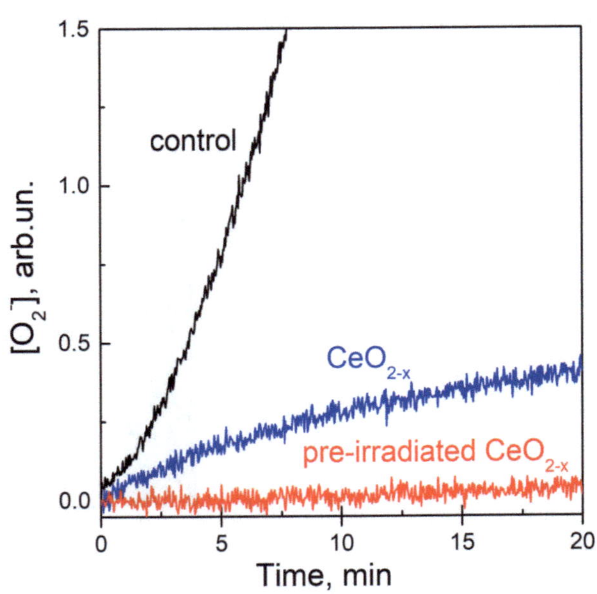

Fig. 4.5 Dynamics of the concentration of superoxide anions (O_2^-) at addition of CeO_{2-x} nanocrystals (2 nm) determined by the fluorescence intensity of adrenochrome (520 nm)

(O_2^-) in aqueous solutions. In general, neutralization of superoxide anions can occur in two possible ways:

$$O_2^- + Ce^{3+} + 2H^+ \rightarrow H_2O_2 + Ce^{4+} \quad (4.1)$$

$$O_2^- + Ce^{4+} \rightarrow O_2 + Ce^{3+} \quad (4.2)$$

The conducted experiments make it possible to give preference to the first mechanism of neutralization of superoxide anions (4.1), since according to this mechanism the increase in the content of Ce^{3+} ions should lead to an increase in the antioxidant activity of nanoparticles, which is observed experimentally. Thus, the study of pre-irradiated nanoparticles made it possible to draw conclusions about the specific mechanism of SOD-like activity of nanoparticles.

It should be noted that the effect of pre-irradiation of nanocrystals on their antioxidant activity was observed, as expected, only during the first hours after irradiation, after which the antioxidant activity of irradiated and non-irradiated nanocrystals was the same. So, no noticeable effect of irradiation on the processes of neutralization of long-lived ROS (such as hydrogen peroxide and hypochlorite anions) was observed, although a similar effect should also occur for small concentrations of these ROS.

4.2 pH Effect on the Redox Properties of CeO_{2-x} Nanocrystals

The type of redox activity of oxide nanocrystals can vary depending on external conditions (in particular, pH), that was previously discussed in a number of publications [11, 12]. It was shown that CeO_{2-x} nanoparticles at high pH values demonstrate both SOD-like (transformation of O_2^- anions into H_2O_2 molecules) and catalase-like activity (decomposition of H_2O_2 into water and oxygen), while at low pH values—only SOD-like activity, which leads to increase in the concentration of hydrogen peroxide as a product. In turn, due to the oxidase- and peroxidase-like activity of CeO_{2-x} nanoparticles at low pH values, the excess hydrogen peroxide obtained as a result of SOD-like activity is used to oxidize biological molecules, which can lead to cell death.

The possible role of nanocrystalline cerium oxide as a dual agent capable of killing cancer cells (which have low pH level) and protecting normal cells (which have higher pH level) from oxidative stress was proposed, in particular, in [13]. The ability of CeO_{2-x} nanocrystals to selectively protect normal (non-cancerous) cells from the effects of ROS is confirmed by a number of biological studies [14] under the conditions of ROS generation by X-ray radiation (which simulates the process of radiotherapy of oncological neoplasms). In some publications, it was shown that CeO_{2-x} nanoparticles in cancer cells demonstrate prooxidant activity, contributing to their destruction [15]. However, other authors observed only the antioxidant activity of CeO_{2-x} nanoparticles in both healthy and cancer cells [16]. Such a discrepancy in experimental results indicates, in particular, the key role of the parameters of specific nanoparticle (size, shape, surface morphology, oxygen stoichiometry, etc.) in the formation of its redox properties.

The type of redox activity of oxide nanocrystals with ions of variable valence (in particular, CeO_{2-x} nanocrystals) directly depends on the characteristics of the environment (such as pH and oxygen concentration), and at certain pH values, a transition from pro- to antioxidant activity of nanocrystals should be observed. This fact can be of key importance when obtaining nanocrystals for the needs of oncology, since the pH of cancerous and healthy cells is different (the pH of healthy cells is 7.3–7.4, and the pH of cancerous cells is 6.2–6.9), and therefore the same nanoparticles themselves can act as prooxidants in cancer cells and protect healthy cells from the action of ROS.

As already was noted in Chap. 3, the processes of decomposition of hydrogen peroxide by nanocrystals of cerium oxide can be either catalase-like (without the formation of hydroxyl radicals) or follow a Fenton-like mechanism (accompanied by the formation of hydroxyl radicals). Decomposition of hydrogen peroxide by Fenton-like mechanism:

$$H_2O_2 + Ce^{3+} \rightarrow Ce^{4+} + OH^- + {}^{\bullet}OH \tag{4.3}$$

can be dangerous for living cells, as it is accompanied by the formation of hydroxyl radicals ($^{\cdot}OH$), the oxidizing ability of which significantly exceeds the oxidizing ability of hydrogen peroxide. Consequently, Fenton-like reaction can lead to peroxidation of lipid membranes and damage to various cell organelles by hydroxyl radicals. Another possible way of interaction of cerium oxide with hydrogen peroxide involves the formation of superoxide anions as an intermediate product:

$$H_2O_2 + Ce^{4+} \rightarrow Ce^{3+} + O_2^- + 2H^+ \tag{4.4}$$

This pathway is less harmful to the cell because superoxide anions can be neutralized by ceria nanocrystals (SOD-like activity of ceria nanocrystals):

$$O_2^- + Ce^{4+} \rightarrow Ce^{3+} + O_2 \tag{4.5}$$

and

$$O_2^- + Ce^{3+} + 2H^+ \rightarrow Ce^{4+} + H_2O_2 \tag{4.6}$$

Therefore, the types of ROS formed during the decomposition of hydrogen peroxide by cerium oxide nanocrystals determine the overall biological activity of these nanoparticles. One of the main factors that control the involvement of a specific reaction of neutralization/formation of ROS is the pH of an aqueous solution of hydrogen peroxide with cerium oxide nanocrystals.

The dynamics of the decomposition of hydrogen peroxide by nanocrystals of cerium oxide in aqueous solutions with different pH was measured using DPPP (diphenyl-1-pyrenylphosphine), a selective sensor for hydrogen peroxide. Upon oxidation to phosphine oxide, the sensor converts to a broadband fluorescent form with a maximum at 380 nm. During the experiment, 980 μL of an aqueous solution of hydrogen peroxide with nanocrystals of cerium oxide taken at different times after the addition of hydrogen peroxide was added to the cuvettes followed by the addition of 20 μL of a DPPP-DMSO solution (1 mM). The resulting solution was kept for 30 min to complete the reaction of the sensor with hydrogen peroxide, and then the fluorescence spectra were recorded. The fluorescence intensity of the sensor was thus proportional to the concentration of hydrogen peroxide at a given time. Figure 4.6 shows the dynamics of the concentration of hydrogen peroxide during the interaction of hydrogen peroxide with cerium oxide nanocrystals at pH = 4, 7.5, and 10. The dynamics consists of two well-defined stages, corresponding to a decrease and a further increase (for pH = 7.5 and 10) of hydrogen peroxide concentration. This specific dynamics can be considered as a result of the combination of catalase-like and SOD-like activity of cerium oxide nanocrystals.

An increase in the pH value leads to a noticeable increase in the rate of decomposition of hydrogen peroxide at the first stage of decomposition.

Interaction of hydrogen peroxide molecules with Ce^{3+}-Vo-Ce^{3+} complexes (or Ce^{4+}-Vo-Ce^{4+} in the oxidized state), which are the most common type of defects on the surface of cerium oxide, ensures the decomposition of hydrogen peroxide into

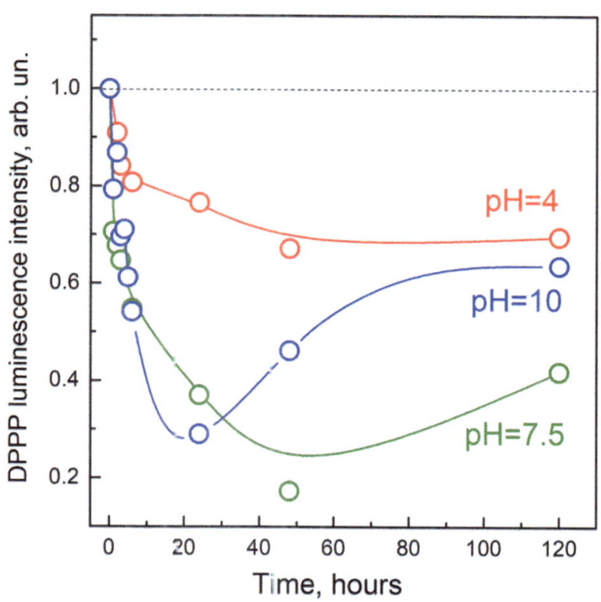

Fig. 4.6 Dynamics of decomposition of hydrogen peroxide by cerium oxide nanocrystals at different pH values, determined by the change in the luminescence intensity of the DPPP sensor ($\lambda = 378$ nm) after the addition of hydrogen peroxide

water and oxygen without the formation of intermediate ROS. At the same time, the interaction with individual Ce^{3+}/Ce^{4+} ions should be accompanied by the formation of ROS. Therefore, in the process of decomposition of hydrogen peroxide, in addition to the formation of oxygen, superoxide anions (O_2^-) can also be formed by reaction (4.4).

Thus, at the second stage of decomposition of hydrogen peroxide, the SOD-like activity of cerium oxide nanocrystals dominates, which leads to the formation of additional hydrogen peroxide, but its content is much less than the initial concentration of hydrogen peroxide.

The formation of superoxide anions during the decomposition of hydrogen peroxide by cerium oxide nanocrystals was studied by changes in the absorption spectra of adrenaline after the addition of hydrogen peroxide. The absorption band at 305 nm corresponds to the absorption of adrenochrome (adrenochrome is a product of the interaction of adrenaline with O_2^- anions). No oxidation of adrenaline to adrenochrome was observed in solutions of adrenaline with hydrogen peroxide without cerium oxide nanocrystals.

The addition of cerium oxide nanocrystals leads to rapid decomposition of hydrogen peroxide according to reaction (4.4), which is accompanied by the formation of superoxide anions (Fig. 4.7a). The content of O_2^- anions formed during the decomposition of hydrogen peroxide depends on pH and is the highest at pH = 10.

Using coumarin as a sensor of hydroxyl radicals, the processes of formation of $^{\bullet}OH$ radicals during the decomposition of hydrogen peroxide by cerium oxide nanocrystals at different pH values were studied. A decrease in pH increases the content of hydroxyl radicals formed during the decomposition of hydrogen peroxide (Fig. 4.7b) in the presence of cerium oxide nanocrystals. At the same time, the neutralization of $^{\bullet}OH$ radicals by cerium oxide nanocrystals can eliminate a large part of the hydroxyl radicals formed during the decomposition of hydrogen peroxide. For cerium oxide nanocrystals at pH = 7, the content of hydroxyl radicals is lower

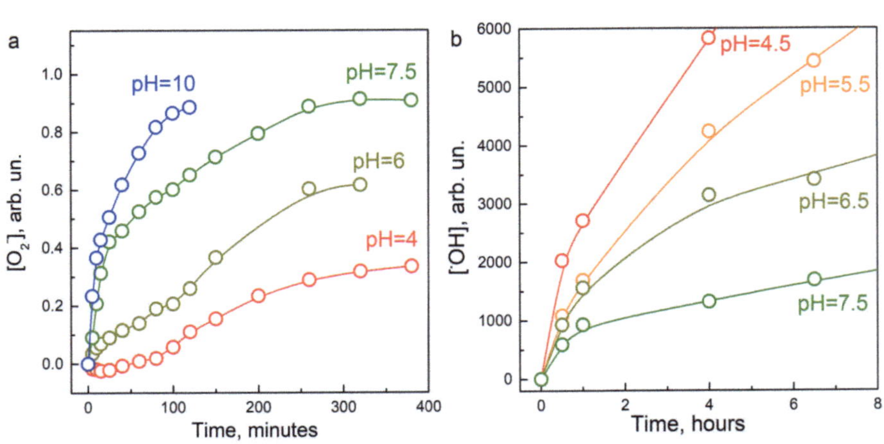

Fig. 4.7 The concentration of superoxide anions (O_2^-) (**a**) and hydroxyl radicals ($^{\bullet}OH$) (**b**) in aqueous solutions of hydrogen peroxide and cerium oxide nanocrystals with different pH values

than in the control water-hydrogen peroxide solution without cerium oxide nanocrystals throughout the hydrogen peroxide decomposition process.

In Chap. 3, two factors were shown that reduce the formation of hydroxyl radicals in the presence of cerium oxide nanocrystals: first, a high content of Ce^{3+}-Vo-Ce^{3+} surface complexes able to decompose hydrogen peroxide without the formation of hydroxyl radicals, and, secondly, a high ability to neutralize hydroxyl radicals. The ability of cerium oxide nanocrystals to neutralize hydroxyl radicals should also be pH-dependent. In the experiment, aqueous solutions of coumarin with cerium oxide nanoparticles at different pH values were irradiated with X-ray radiation ($U = 30$ kV, $I = 20$ mA), which led to the radiolysis of water with the formation of hydroxyl radicals. The fluorescence of 7-hydroxycoumarin was used to estimate the content of hydroxyl radicals in solutions. It was shown that a decrease in pH reduces the ability of cerium oxide nanocrystals to neutralize hydroxyl radicals (Fig. 4.8). Thus, cerium oxide nanocrystals can both decompose hydrogen peroxide with the formation of hydroxyl radicals and neutralize radicals obtained in the process of peroxide decomposition, and the final prooxidant or antioxidant effect of cerium oxide nanocrystals must be the result of the combined action of the processes of formation and neutralization of $^{\bullet}OH$ radicals at certain pH values.

In order to reveal the dominant type of redox activity of cerium oxide nanocrystals at different pH values, the change in $^{\bullet}OH$ content in aqueous solutions of hydrogen peroxide with and without cerium oxide nanocrystals was compared. It is known that the generation of $^{\bullet}OH$ radicals from hydrogen peroxide is possible even without cerium oxide nanocrystals due to the presence of small amounts of

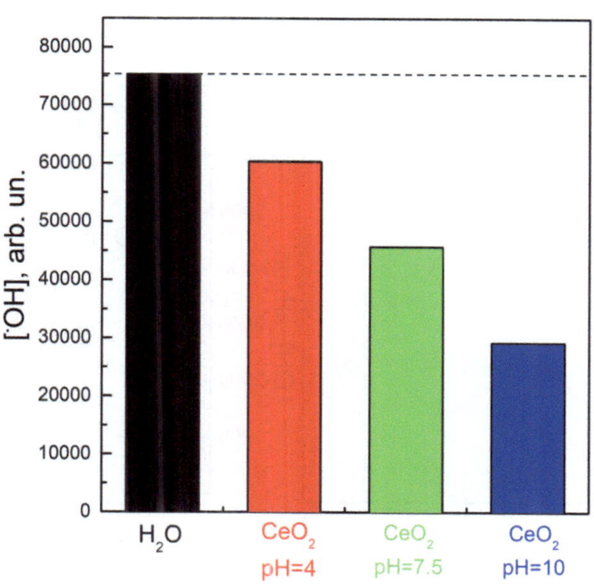

Fig. 4.8 Neutralization of hydroxyl radicals ($^{\bullet}OH$) formed at X-ray irradiation in aqueous solutions by cerium oxide nanocrystals

uncontrolled impurities (dissolved ions, etc.). A comparison between samples with and without cerium oxide nanocrystals shows that, despite the ability of nanocrystals to form hydroxyl radicals, the neutralization of hydroxyl radicals by cerium oxide nanocrystals prevails at almost all pH values (Fig. 4.9a–d).

At the same time, the difference between the content of $^{\bullet}OH$ radicals formed in aqueous solutions with and without cerium oxide nanocrystals decreases with pH from 7.5 to 4.5. Thus, at a lower pH value, the protection provided by cerium oxide nanocrystals against $^{\bullet}OH$ radicals decreases, and at pH = 4.5, at the first stage of hydrogen peroxide decomposition, the prooxidant properties of cerium oxide nanocrystals even prevail over antioxidant properties (Fig. 4.9a).

Monitoring the Ce^{3+}/Ce^{4+} ratio in cerium oxide nanocrystals using luminescent methods can provide a deeper insight into the mechanism of formation and neutralization of specific types of ROS during the decomposition of hydrogen peroxide at different pH values.

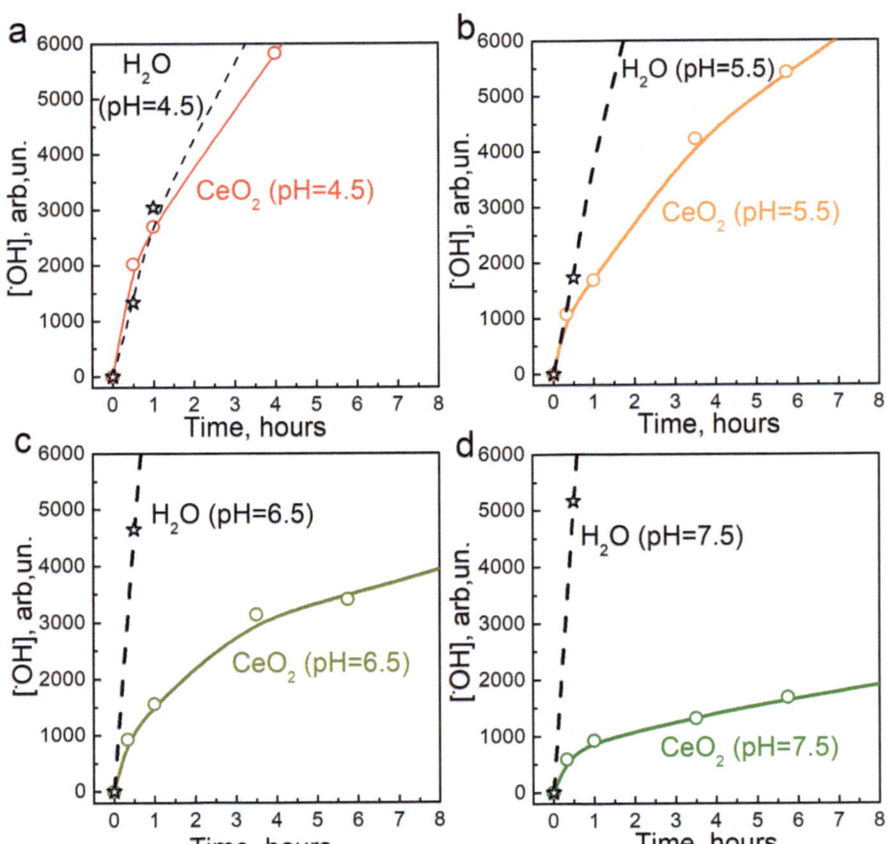

Fig. 4.9 Comparison of the content of hydroxyl radicals ($^{\bullet}OH$) in aqueous solutions of hydrogen peroxide with different pH values (**a–d**) with and without cerium oxide nanocrystals, which was determined by the change in 7-hydroxycoumarin luminescence ($\lambda = 460$ nm)

Hydrogen peroxide can act both as an oxidant and as a reducing agent forming different types of ROS ($^{\cdot}OH$ or O_2^-) during interaction with active surface sites ($Ce^{3+} \rightarrow Ce^{4+}$ ions or Ce^{3+}-Vo-Ce^{3+} complexes).

Figure 4.10 shows the change in Ce^{3+} luminescence intensity after adding hydrogen peroxide at different pH values. At pH = 7.5 and pH = 6, the addition of hydrogen peroxide leads to fast $Ce^{3+} \rightarrow Ce^{4+}$ oxidation, at pH = 10 to fast $Ce^{3+} \rightarrow Ce^{4+}$ reduction, and at pH = 4, the Ce^{3+}/Ce^{4+} ratio remains the same as before an addition of hydrogen peroxide. These results pave the way to understanding the entire mechanism of pH-dependent formation of $^{\cdot}OH$ radicals or superoxide anions during the decomposition of hydrogen peroxide.

As can be seen from Fig. 4.7, the way of decomposition of hydrogen peroxide by cerium oxide nanocrystals (corresponding to reaction (4.3) or reaction (4.4)) can be controlled by pH. An increase in the content of OH^- ions (pH = 10) can stimulate the decomposition according to reaction (4.4), since this reaction leads to an increase in the content of H^+ ions, compensating for the initial excess of OH^- ions. This reaction should be accompanied by the formation of superoxide anions and the reduction of Ce^{4+} ions to Ce^{3+}, which is consistent with the increase in the content of O_2^- ions and the increase in the intensity of Ce^{3+} luminescence in Fig. 4.7a and Fig. 4.10, respectively.

At weakly acidic pH values (pH = 6), an excess of H^+ ions, in turn, provides more efficient decomposition of hydrogen peroxide according to reaction (4.3), which leads to the formation of OH^- anions. Reaction (4.3) involves the formation of $^{\cdot}OH$ radicals and the oxidation of Ce^{3+} ions to Ce^{4+}, which explains the higher content of

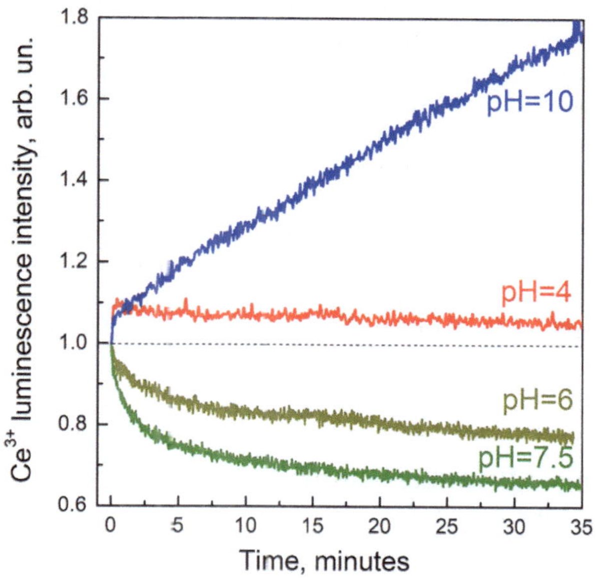

Fig. 4.10 Dynamics of Ce^{3+} luminescence intensity after the addition of hydrogen peroxide to aqueous solutions of nanoceria with different pH values

$^{\cdot}OH$ radicals at lower pH values and the decrease in the intensity of Ce^{3+} luminescence, which can be seen in Fig. 4.7b and Fig. 4.10, respectively. For pH = 4, the concentration of $^{\cdot}OH$ radicals is high, but the reversibility of $Ce^{4+} \leftrightarrow Ce^{3+}$ transition is disrupted due to the high content of H^+ ions [17], so the Ce^{3+}/Ce^{4+} ratio (and, therefore, the Ce^{3+} luminescence intensity) remains almost constant during hydrogen peroxide decomposition.

The resulting dynamics of decomposition of hydrogen peroxide (Fig. 4.6) by cerium oxide nanocrystals at a certain pH value is the result of its catalase- and SOD-like activity together with the ability to form superoxide anions. The higher the pH value, the higher the content of O_2^- anions, which are formed during the decomposition of hydrogen peroxide. At a certain concentration of superoxide anions in the aqueous solution, the SOD-like activity of nanoceria begins to dominate (reactions (4.5) and (4.6)), which leads to the formation of additional hydrogen peroxide, which can be seen in Fig. 4.6 to increase the fluorescence intensity of the hydrogen peroxide sensor. Since the content of superoxide anions at pH = 10 is the highest, at this pH value, the most significant increase in the content of hydrogen peroxide is observed. On the contrary, for pH = 4, the formation of additional hydrogen peroxide was not observed, since the decomposition of hydrogen peroxide at this pH occurs mainly according to reaction (4.3) and is not accompanied by the formation of superoxide. Thus, a change in the pH of an aqueous solution of cerium oxide nanocrystals changes not only the rate, but also the mechanism of decomposition of hydrogen peroxide.

Therefore, both the rate and specific mechanisms of decomposition of hydrogen peroxide by cerium oxide nanocrystals depend on the pH value of aqueous solutions of hydrogen peroxide with cerium oxide nanocrystals. The higher the pH value, the faster the decomposition of hydrogen peroxide. Two different ways of decomposition of hydrogen peroxide by cerium oxide nanocrystals were revealed: the first one with the formation of superoxide anions, which prevails at high pH values, and the second with the formation of hydroxyl radicals, which dominates at low pH values. The high ability of cerium oxide nanocrystals to neutralize $^{\bullet}OH$ radicals leads to almost complete removal of hydroxyl radicals, which are formed during the decomposition of hydrogen peroxide at neutral and weakly acidic pH. At the same time, at pH = 4.5, the prooxidant activity of cerium oxide nanocrystals is revealed, which may be associated with inhibition of oxidation-reduction processes of Ce^{3+}/Ce^{4+} ions on the surface of nanocrystals, which leads to low efficiency of $^{\bullet}OH$ radical neutralization.

4.3 Hydrogen Peroxide Sensors Based on Cerium Oxide Nanocrystals

The change in the intensity of Ce^{3+} luminescence of CeO_2 nanocrystals upon addition of oxidants (in particular, ROS) opens up the possibility of using these materials not only as antioxidants, but also as sensors of the concentration of the corresponding oxidants in aqueous solutions. In particular, the development of sensors for detecting the concentration of the most widespread form of ROS, hydrogen peroxide (H_2O_2) is of the greatest interest from both a biological and a technical point of view.

Today, direct and indirect methods of H_2O_2 sensing are represented by fluorimetry [18], chemiluminescence [19], and spectrophotometry [20]. In recent years, a large number of new sensors have been developed (in particular, based on Prussian blue (PB) [21, 22], heme proteins [23], carbon nanotubes [24], and various metal nanoparticles [25]), improving the sensitivity and efficiency of hydrogen peroxide detection methods.

Luminescent sensors were proven to be the most reliable sensors for detecting hydrogen peroxide in living cells [26–28]. For example, one of the most common sensors is fluorescein, the action of which is based on the transition of the non-fluorescent form of fluorescein into the fluorescent form (dichlorofluorescein) when hydrogen peroxide is added [29]. Dichlorofluorescein assay is recognized as one of the most important in nanotoxicology for the quantification of oxidative stress caused by various factors (including external action, penetration of nanoparticles, etc.). However, this assay has a number of limitations due to low selectivity and the probable ability of dichlorofluorescein to generate ROS. Enzyme-based hydrogen peroxide biosensors show high selectivity, but they have a limited lifetime.

The limitations inherent to sensors based on organic molecules make it necessary to develop hydrogen peroxide sensors that would be stable in aqueous solutions and biological fluids for a long time, and possess reversible redox properties for their multiple use.

The unique redox properties of cerium oxide nanoparticles shown in a number of works allow this material to be used not only for neutralizing ROS, but also for effective detection of ROS (including hydrogen peroxide) in biological samples. In [30], an amperometric hydrogen peroxide sensor consisting of a working electrode covered with 3–5 nm cerium oxide nanoparticles, an Ag/AgCl reference electrode, and a gold counter electrode on a glass substrate was proposed. The high sensitivity and linear response observed in the range from 1 to 50 μM allowed the authors to confirm the prospects of this sensor for detecting small concentrations of hydrogen peroxide.

The presence of Ce^{3+} luminescence in cerium oxide nanocrystals, which was shown in the previous sections, opens up the possibility of detecting hydrogen peroxide in aqueous solutions using optical spectroscopy methods, which in many applications (primarily biological ones) are much more convenient than amperometric methods [30].

In order to quantify the effect of the content of hydrogen peroxide in aqueous solutions on the intensity of Ce^{3+} luminescence in cerium oxide nanocrystals, various concentrations of hydrogen peroxide were added to colloidal solutions of cerium oxide nanocrystals. Luminescence spectra of cerium oxide nanocrystals were measured before and 10 min after the addition of hydrogen peroxide. To prevent the effect of inhomogeneous distribution of the oxidant when it was added to a cuvette with a colloidal solution of cerium oxide, the cuvette was placed on a magnetic stirrer.

In the Figs. 4.11a and 4.12a, luminescence spectra of 2 and 10 nm cerium oxide nanocrystals depending on the concentration of hydrogen peroxide in cerium oxide colloidal solutions are shown. The Ce^{3+} luminescence intensity decreases monotonically with increasing H_2O_2 concentration from 2.5 to 200 mM for both 2 and 10 nm

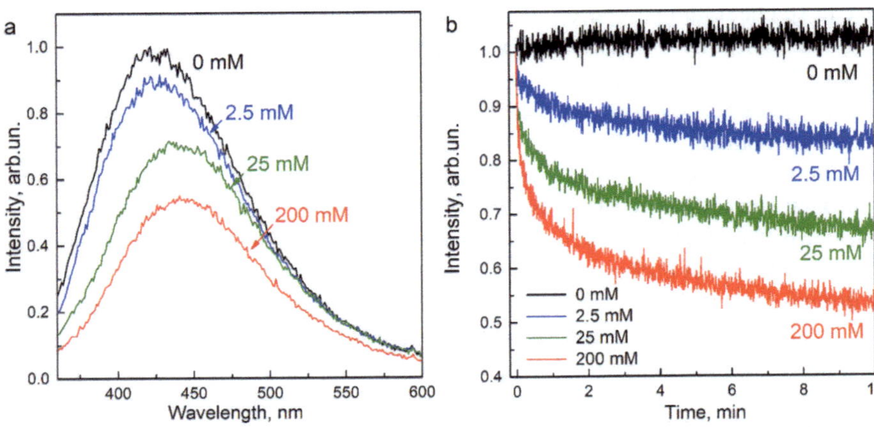

Fig. 4.11 (**a**) Luminescence spectra of colloidal solutions of 2 nm CeO_{2-x} nanocrystals ($\lambda_{exc} = 325$ nm) depending on the concentration of hydrogen peroxide in an aqueous solution. (**b**) Time dependences of Ce^{3+} luminescence (430 nm) after adding different concentrations of hydrogen peroxide to colloidal solutions of 2 nm nanocrystals. (Reprinted with permission from *Optical Materials*, vol. 85, Y.Malyukin, V. Seminko, P. Maksimchuk, E. Okrushko, O. Sedyh and Y. Zorenko, Hydrogen peroxide sensing using Ce^{3+} luminescence of cerium oxide (CeO_{2-x}) nanoparticles, pp. 303–307, 2018. Copyright 2018 Elsevier)

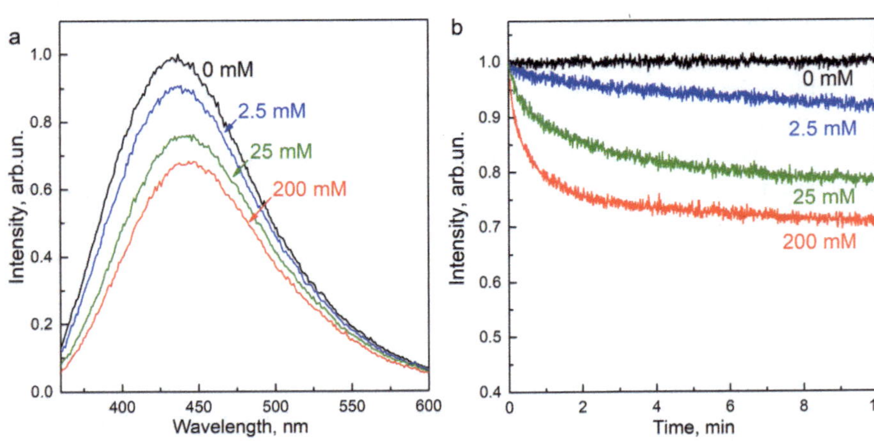

Fig. 4.12 (**a**) Luminescence spectra of colloidal solutions of 10 nm CeO_{2-x} nanocrystals ($\lambda_{exc} = 325$ nm) depending on the concentration of hydrogen peroxide in an aqueous solution. (**b**) Time dependences of Ce^{3+} luminescence (440 nm) after adding different concentrations of hydrogen peroxide to colloidal solutions of 10 nm nanocrystals. (Reprinted with permission from *Optical Materials*, vol. 85, Y.Malyukin, V. Seminko, P. Maksimchuk, E. Okrushko, O. Sedyh and Y. Zorenko, Hydrogen peroxide sensing using Ce^{3+} luminescence of cerium oxide (CeO_{2-x}) nanoparticles, pp. 303–307, 2018. Copyright 2018 Elsevier)

nanocrystals. The decrease in luminescence intensity is accompanied by a slight shift of the band maximum from 425 to 440 nm for 2 nm nanoparticles and from 430 to 445 nm for 10 nm nanoparticles. The time dependences of Ce^{3+} luminescence obtained at 430 nm for 2 nm nanoparticles and at 440 nm for 10 nm cerium oxide nanocrystals are shown in Fig. 4.11b and Fig. 4.12b, respectively.

The time dependences can be divided into two stages: the stage of fast decrease of Ce^{3+} luminescence (1–2 min) and the stage of relatively slow decrease of Ce^{3+} luminescence. The intensity of Ce^{3+} luminescence after adding hydrogen peroxide directly depends on the concentration of H_2O_2 in the resulting solution.

In Fig. 4.13, the dependence of the intensity of Ce^{3+} luminescence on the concentration of hydrogen peroxide (10 min after addition) for colloidal solutions of 2 and 10 nm cerium oxide nanoparticles is shown. In the semilogarithmic scale, both dependences are close to linear in the entire range of concentrations. An increase in the concentration of hydrogen peroxide from 1.25 to 200 mM leads to a decrease in the intensity of Ce^{3+} luminescence by 1.7 and 1.4 times for cerium oxide nanoparticles with a size of 2 nm and 10 nm, respectively.

Rate of recovery of initial luminescent signal after interaction with H_2O_2 is one of the main parameters of hydrogen peroxide sensors. As was mentioned in Chap. 3, fast $Ce^{3+} \rightarrow Ce^{4+}$ oxidation after H_2O_2 addition is followed by very slow recovery of the initial Ce^{3+}/Ce^{4+} ratio in ceria nanocrystals. The process of recovery of initial Ce^{3+} content in 2 nm nanoceria can last for more than 120 h at room temperature (Fig. 4.14a). The duration of this process, however, decreases sufficiently at higher

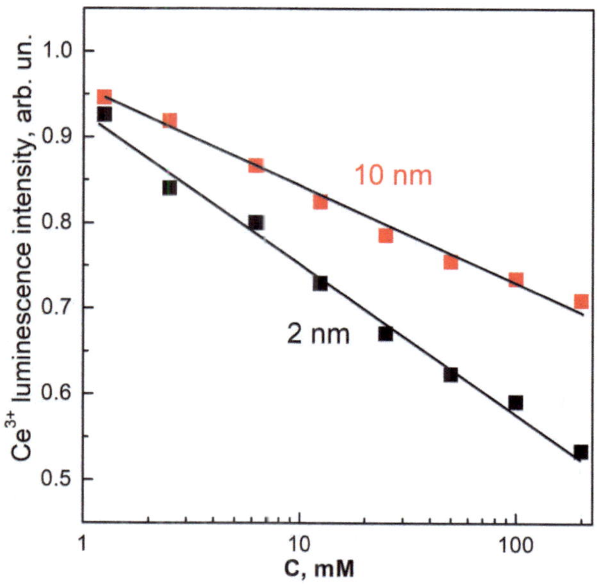

Fig. 4.13 Dependence of Ce^{3+} luminescence intensity on hydrogen peroxide concentration for colloidal solutions of 2 and 10 nm CeO_{2-x} nanocrystals (10 min after H_2O_2 addition). (Reprinted with permission from *Optical Materials*, vol. 85, Y.Malyukin, V. Seminko, P. Maksimchuk, E. Okrushko, O. Sedyh and Y. Zorenko, Hydrogen peroxide sensing using Ce^{3+} luminescence of cerium oxide (CeO_{2-x}) nanoparticles, pp. 303–307, 2018. Copyright 2018 Elsevier)

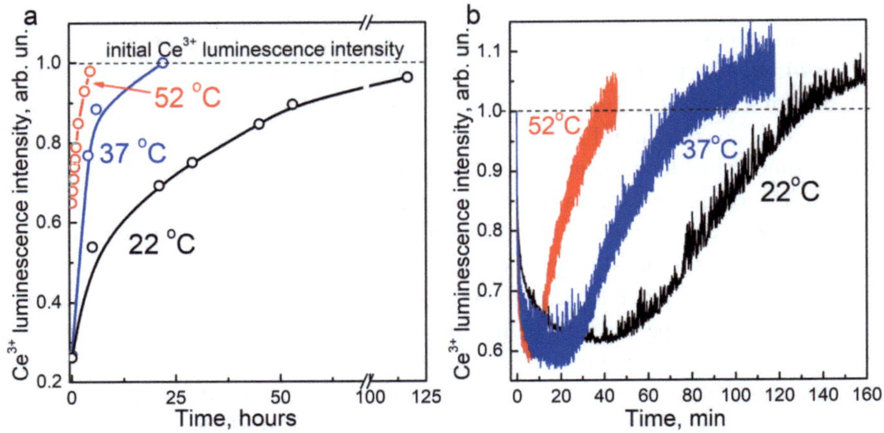

Fig. 4.14 Dynamics of recovery of the initial intensity of Ce^{3+} luminescence ($Ce^{4+} \rightarrow Ce^{3+}$) after H_2O_2 addition to colloidal solutions of CeO_{2-x} nanocrystals (2 nm) at temperatures of 22, 37, and 52 °C: (**a**) in the absence of continuous laser irradiation; (**b**) in the presence of continuous laser irradiation (325 nm)

temperatures (from 120 h at 22 °C to 22 h at 37 °C, and to 4.5 h at 52 °C). The recovery rate can be increased even more by UV irradiation with $\lambda = 325$ nm (Fig. 4.14b). This irradiation leads to $O^{2-} \rightarrow Ce^{4+}$ electron transfer stimulating $Ce^{4+} \rightarrow Ce^{3+}$ reduction, and so, favors the processes of recovery of Ce^{3+} luminescence intensity. So, at 22 °C, recovery of the initial Ce^{3+}/Ce^{4+} ratio takes about 120 h without irradiation and only 90 min with continuous irradiation; at 37 °C, it takes 24 h without irradiation and 60 min with continuous irradiation, and at 52 °C, it takes 5 h without irradiation and 30 min with continuous irradiation.

No influence of laser irradiation on the processes of recovery of Ce^{3+} luminescence was observed for laser wavelengths that did not fall into the spectral range corresponding to the charge-transfer absorption band ($O^{2-} \rightarrow Ce^{4+}$) (Fig. 4.15a). As can be seen from Fig. 4.15b, the increase of recovery rate of initial Ce^{3+} luminescence intensity is observed for 325 nm continuous laser irradiation, while at 530 nm irradiation, the dynamics of recovery of Ce^{3+} luminescence intensity was the same as without any irradiation, despite the fact that the power of the 530 nm laser diode (50 mW) was even higher than that for He-Cd laser 325 nm (6 mW).

Figure 4.16 shows the dynamics of recovery of the initial intensity of Ce^{3+} luminescence ($Ce^{4+} \rightarrow Ce^{3+}$) after adding hydrogen peroxide during irradiation with a 325 nm laser under the conditions of varying the intensity of irradiation. An increase in the intensity of irradiation from 0.3 to 1.2 W/cm^2 (which corresponds to an increase in the photon flux from $0.5 \bullet 10^{18}$ to $2 \bullet 10^{18}$ photons/s•cm^2) leads to an increase of recovery rate of initial Ce^{3+} luminescence intensity. So, both temperature and additional irradiation can increase the rate of recovery of the initial luminescent signal of CeO_{2-x} hydrogen peroxide sensors.

Oxidation and photoreduction cycles of CeO_{2-x} nanocrystals can be repeated many times, while the luminescence spectrum and oxidation-reduction dynamics remain unchanged (Fig. 4.17).

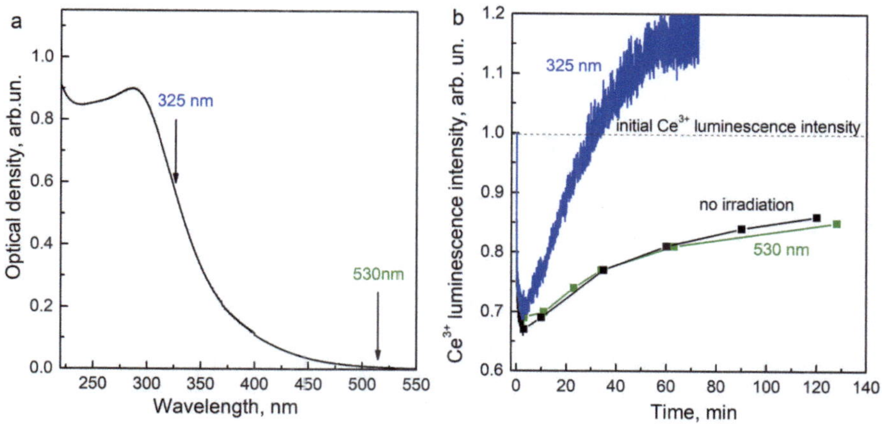

Fig. 4.15 (**a**) Absorption spectrum of colloidal solutions of CeO_{2-x} nanocrystals (2 nm); (**b**) dynamics of recovery of the initial intensity of Ce^{3+} luminescence ($Ce^{4+} \rightarrow Ce^{3+}$) after adding hydrogen peroxide to colloidal solutions of CeO_{2-x} nanocrystals (2 nm) in the presence of continuous laser irradiation with different wavelengths (325 and 530 nm), $t = 52$ °C

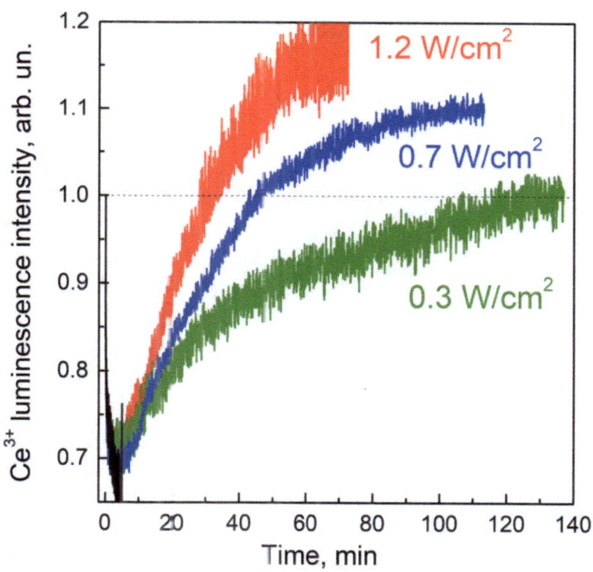

Fig. 4.16 Dynamics of Ce^{3+} luminescence recovery ($Ce^{4+} \rightarrow Ce^{3+}$) after adding hydrogen peroxide to colloidal solutions of CeO_{2-x} nanocrystals (2 nm) at different intensities of continuous laser irradiation (325 nm), $t = 52$ °C

Hydrogen peroxide sensors can be obtained not only using CeO_{2-x}, but also using CeO_{2-x}:Eu^{3+} nanocrystals. As for Ce^{3+} luminescence, Eu^{3+} luminescence of CeO_{2-x}:Eu^{3+} nanocrystals is quenched by H_2O_2 addition (Fig. 4.18a, b). This quenching cannot be caused by Eu^{3+} oxidation, so, most probably, this effect is due to energy transfer between Eu^{3+} ions by H_2O_2 and/or hydroxyl groups formed on the surface of

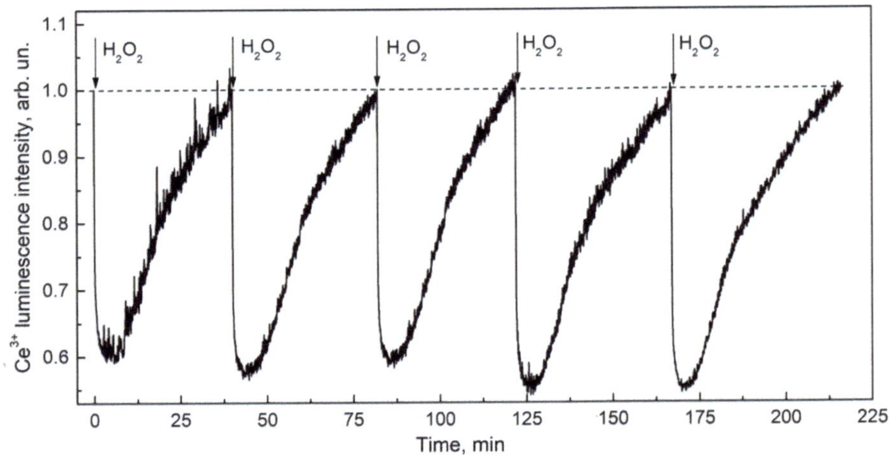

Fig. 4.17 Dynamics of Ce^{3+} luminescence intensity of CeO_{2-x} nanocrystals (2 nm) at multiple H_2O_2 addition and continuous UV irradiation. $t = 52$ °C. (Reprinted with permission from *Journal of Physical Chemistry C*, vol. 127, V. Seminko, P. Maksimchuk, V. Klochkov, Y. Neuhodov, L. Demchenko, S. Yefimova, Reversible CeO_{2-x} and CeO_{2-x}:Eu^{3+} Luminescent Hydrogen Peroxide Sensors with Recovery Rates Controlled by Temperature and UV Irradiation, pp. 10662–10669, 2023. Copyright 2023 American Chemical Society)

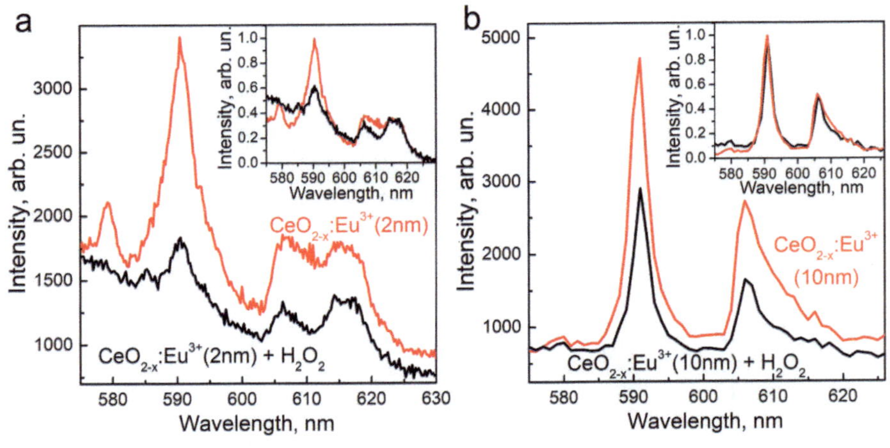

Fig. 4.18 Hydrogen peroxide sensing using quenching of and Eu^{3+} luminescence of CeO_{2-x}:Eu^{3+} (**a**—2 nm, **b**—10 nm) colloidal nanocrystals. Normalized luminescence spectra are shown in the insets. (Reprinted with permission from *Journal of Physical Chemistry C*, vol. 127, V. Seminko, P. Maksimchuk, V. Klochkov, Y. Neuhodov, L. Demchenko, S. Yefimova, Reversible CeO_{2-x} and CeO_{2-x}:Eu^{3+} Luminescent Hydrogen Peroxide Sensors with Recovery Rates Controlled by Temperature and UV Irradiation, pp. 10662–10669, 2023. Copyright 2023 American Chemical Society)

nanocrystals as a result of H_2O_2 adsorption and following decomposition. Eu^{3+} luminescence quenching by hydroxyl groups was previously shown for SnO_2:Eu^{3+} nanocrystals [31]. It should be noted that H_2O_2 sensing using Eu^{3+} luminescence quenching can provide some advances over H_2O_2 sensing using Ce^{3+} luminescence as red emission of europium ions has lower overlap with autofluorescence of biological tissues that determine wide use of Eu^{3+}-doped nanocrystals for biological applications. The potential use of Eu^{3+}-doped nanoceria for H_2O_2 and glucose sensing was at first proposed in [32], where the dependence of Eu^{3+} luminescence intensity on the H_2O_2 concentration was shown. These sensors were successfully used for H_2O_2 real-time detection in bacterial cell cultures confirming the high potential of these materials for real-time in situ H_2O_2 monitoring in biological objects.

In Fig. 4.18a, b, the red parts of the luminescence spectra of 2 nm and 10 nm CeO_{2-x}:Eu^{3+} nanocrystals are shown. For 2 nm nanocrystals besides $^5D_0 \rightarrow {}^7F_1$ (591 nm) and $^5D_0 \rightarrow {}^7F_2$ (605–615 nm), also $^5D_0 \rightarrow {}^7F_0$ (580 nm) Eu^{3+} luminescence bands are observed. Due to proximity of large number of Eu^{3+} ions to the surface of NP and high oxygen deficiency of 2 nm CeO_{2-x}:Eu^{3+} nanocrystals, all Eu^{3+} luminescence bands for 2 nm nanocrystals are wider as compared to 10 nm nanocrystals. The intensity of Eu^{3+} luminescence decreases in 1.9 times after addition of 40 mM of H_2O_2 for 2 nm CeO_{2-x}:Eu^{3+} nanocrystals (Fig. 4.18a) and in 1.6 times for 10 nm CeO_{2-x}:Eu^{3+} nanocrystals (Fig. 4.18b). Higher quenching for 2 nm nanocrystals can be related to higher quota of Eu^{3+} ions accessible for quencher compared to 10 nm nanocrystals.

The spectra normalized by intensities of $^5D_0 \rightarrow {}^7F_2$ luminescence bands are shown in the insets of Fig. 4.18a, b. While for 10 nm nanocrystals, the ratio between $^5D_0 \rightarrow {}^7F_1$ and $^5D_0 \rightarrow {}^7F_2$ luminescence bands remain the same before and after H_2O_2 addition, and for 2 nm nanoparticles, the redistribution of intensities of Eu^{3+} luminescence bands is observed. It is known that the ratio between $^5D_0 \rightarrow {}^7F_1$ and $^5D_0 \rightarrow {}^7F_2$ luminescence bands can provide information on the symmetry of Eu^{3+} ions in nanocrystals as $^5D_0 \rightarrow {}^7F_1$ transition is symmetry insensitive (magnetodipole), while $^5D_0 \rightarrow {}^7F_2$ transition is highly sensitive to symmetry (electrodipole). For 10 nm nanocrystals, the $^5D_0 \rightarrow {}^7F_1/{}^5D_0 \rightarrow {}^7F_2$ ratio both before and after H_2O_2 addition is ~1.75, while for 2 nm, this ratio is ~1.85 before H_2O_2 addition and ~1.4 after H_2O_2 addition. So, for 2 nm nanocrystals, H_2O_2 addition besides quenching of Eu^{3+} luminescence also decreases the symmetry of Eu^{3+} ions. The details of H_2O_2 interaction with the surface of nanoceria leading to the shift of Ce^{3+} luminescence maxima and redistribution of Eu^{3+} luminescence bands will be discussed below.

The slow reversibility of Eu^{3+} luminescence intensity after H_2O_2 addition was mentioned in [32]. These results agree with our own studies on the dynamics of Ce^{3+} luminescence intensity after H_2O_2 addition. We have done the same experiments as for CeO_{2-x} nanocrystals also for CeO_{2-x}:Eu^{3+} nanocrystals monitoring the recovery of the sensors using both Ce^{3+} (430 nm) and Eu^{3+} (591 nm) luminescence (Fig. 4.19a, b). The clear correlation between the processes of recovery of Ce^{3+} and Eu^{3+} luminescence shows that these processes are intrinsically connected. Without continuous UV irradiation, the recovery of both Ce^{3+} and Eu^{3+} luminescence bands needs about 4.5 h at 52 °C (Fig. 4.19a), while at 325 nm UV irradiation, this process lasts only about 1 h (Fig. 4.19b). Interestingly, the time dependence of Eu^{3+} luminescence in the presence of UV irradiation differs sufficiently from the same dependence without continuous UV irradiation as after recovery of initial intensity of Eu^{3+} luminescence, this intensity starts to decrease again. This process can be related to $O^{2-} \rightarrow Eu^{3+}$ charge transfer stimulated by UV irradiation that competes with the

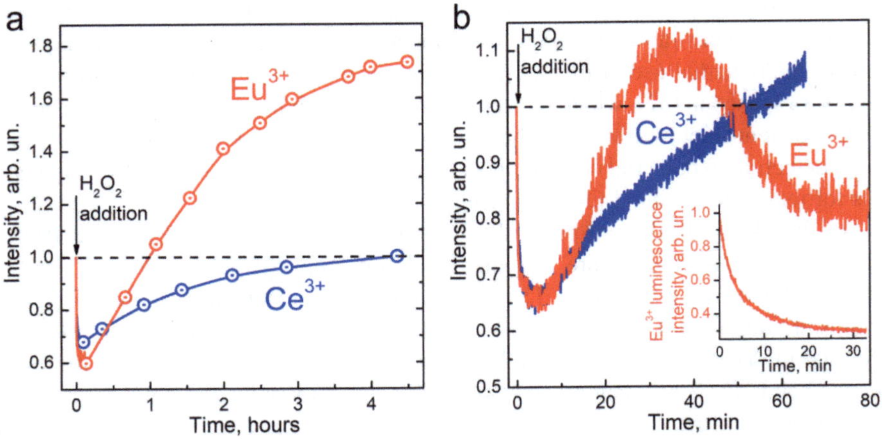

Fig. 4.19 Recovery of initial luminescence intensity of 2 nm CeO_{2-x}:Eu^{3+} (**a, b**) colloidal nanocrystals after H_2O_2 addition without irradiation (**a**) and with continuous UV irradiation (**b**). Inset–decrease of Eu^{3+} luminescence intensity of CeO_{2-x}:Eu^{3+} NCs due to UV-induced photoreduction (without H_2O_2 addition). $t = 52$ °C. (Reprinted with permission from *Journal of Physical Chemistry C*, vol. 127, V. Seminko, P. Maksimchuk, V. Klochkov, Y. Neuhodov, L. Demchenko, S. Yefimova, Reversible CeO_{2-x} and CeO_{2-x}:Eu^{3+} Luminescent Hydrogen Peroxide Sensors with Recovery Rates Controlled by Temperature and UV Irradiation, pp. 10662–10669, 2023. Copyright 2023 American Chemical Society)

process of recovery of initial intensity of Eu^{3+} luminescence after H_2O_2 decomposition. In the inset to Fig. 4.19b, the dependence of Eu^{3+} luminescence intensity in CeO_{2-x}:Eu^{3+} nanocrystals at 325 nm irradiation in the absence of H_2O_2 is shown which is similar to the last stage in the time dependences of Eu^{3+} luminescence observed after H_2O_2 decomposition. This stage can be avoided simply by switching off UV irradiation after full recovery of Eu^{3+} luminescence intensity and is shown here only for the sake of completeness.

As the process of H_2O_2 decomposition is over, the recovery of initial Ce^{3+} and Eu^{3+} luminescence intensity takes place. The processes of $Ce^{4+} \rightarrow Ce^{3+}$ recovery can be directly influenced by 325 nm UV irradiation as a result of $O^{2-} \rightarrow Ce^{4+}$ electron transfer. Also UV can stimulate removing the remaining hydroxyl groups from the surface of nanoparticle increasing the rate of recovery of initial Eu^{3+} luminescence intensity (however, this effect is partially hided by $O^{2-} \rightarrow Eu^{3+}$ charge transfer). In any way, controlling the processes of recovery of undoped and doped nanoceria H_2O_2 luminescence sensors can be useful for applying these sensors in the vast fields of biological, medical, and other techniques.

References

1. Vatanparast M, Saedi L (2018) Sonochemical-assisted synthesis and characterization of CeO_2 nanoparticles and its photocatalytic properties. J Mater Sci Mater Electron 29:7107–7113. https://doi.org/10.1007/s10854-018-8698-8

2. Yuan S, Xu B, Zhang Q, Liu S, Xie J, Zhang M, Ohno T (2018) Development of the visible-light response of CeO_{2-x} with a high Ce^{3+} content and its photocatalytic properties. ChemCatChem 10:1267–1271. https://doi.org/10.1002/cctc.201701767
3. Zhong JB, Ma D, He XY, Li JZ, Chen YQ (2010) Preparation, characterization and photocatalytic performance of $TiO_2/Ce_xZr_{1-x}O_2$ toward the oxidation of gaseous benzene. Appl Surf Sci 256:2859–2862. https://doi.org/10.1016/j.apsusc.2009.11.040
4. Li FB, Li XZ, Hou MF, Cheah KW, Choy WCH (2005) Enhanced photocatalytic activity of Ce^{3+}–TiO_2 for 2-mercaptobenzothiazole degradation in aqueous suspension for odour control. Appl Catal A Gen 285:181–189. https://doi.org/10.1016/j.apcata.2005.02.025
5. Fiorenza R, Bellardita M, D'Urso L, Compagnini G, Palmisano L, Scirè S (2016) Au/TiO_2–CeO_2 catalysts for photocatalytic water splitting and VOCs oxidation reactions. Catalysts 6:121. https://doi.org/10.3390/catal6080121
6. Plata JJ, Márquez AM, Sanz JF (2013) Electron mobility via polaron hopping in bulk ceria: a first-principles study. J Phys Chem C 117:14502–14509. https://doi.org/10.1021/jp402594x
7. Castleton CW, Lee AL, Kullgren J, Hermansson K (2014) Description of polarons in ceria using density functional theory. J Phys Conf Ser 526:012002. https://doi.org/10.1088/1742-6596/526/1/012002
8. Mochizuki S, Fujishiro F (2009) The photoluminescence properties and reversible photoinduced spectral change of CeO_2 bulk, film and nanocrystals. Phys Status Solidi (b) 246:2320–2328. https://doi.org/10.1002/pssb.200844419
9. Mochizuki S, Fujishiro F, Minami S (2005) Photoluminescence and reversible photo-induced spectral change of $SrTiO_3$. J Phys Condens Matter 17:923. https://doi.org/10.1088/0953-8984/17/6/011
10. Trevisanutto PE, Sushko PV, Shluger AL, Beck KM, Henyk M, Joly AG, Hess WP (2005) A mechanism of photo-induced desorption of oxygen atoms from MgO nanocrystals. Surf Sci 593:210–220. https://doi.org/10.1016/j.susc.2005.06.063
11. Shcherbakov AB, Zholobak NM, Spivak NY, Ivanov VK (2014) Advances and prospects of using nanocrystalline ceria in cancer theranostics. Russ J Inorg Chem 59:1556–1575. https://doi.org/10.1134/S003602361413004X
12. Alpaslan E, Yazici H, Golshan NH, Ziemer KS, Webster TJ (2015) pH-dependent activity of dextran-coated cerium oxide nanoparticles on prohibiting osteosarcoma cell proliferation. ACS Biomater Sci Eng 1:1096–1103. https://doi.org/10.1021/acsbiomaterials.5b00194
13. Xu C, Lin Y, Wang J, Wu L, Wei W, Ren J, Qu X (2013) Nanoceria-triggered synergetic drug release based on CeO_2-capped mesoporous silica host–guest interactions and switchable enzymatic activity and cellular effects of CeO_2. Adv Healthc Mater 2:1591–1599. https://doi.org/10.1002/adhm.201200464
14. Colon J, Herrera L, Smith J, Patil S, Komanski C, Kupelian P, Baker CH (2009) Protection from radiation-induced pneumonitis using cerium oxide nanoparticles. Nanomed Nanotechnol Biol Med 5:225–231. https://doi.org/10.1016/j.nano.2008.10.003
15. Tapeinos C, Battaglini M, Prato M, La Rosa G, Scarpellini A, Ciofani G (2018) CeO_2 nanoparticles-loaded pH-responsive microparticles with antitumoral properties as therapeutic modulators for osteosarcoma. ACS Omega 3:8952–8962. https://doi.org/10.1021/acsomega.8b01060
16. Rubio L, Marcos R, Hernández A (2018) Nanoceria acts as antioxidant in tumoral and transformed cells. Chem Biol Interact 291:7–15. https://doi.org/10.1016/j.cbi.2018.06.002
17. Perez JM, Asati A, Nath S, Kaittanis C (2008) Synthesis of biocompatible dextran-coated nanoceria with pH-dependent antioxidant properties. Small 4:552–556. https://doi.org/10.1002/smll.200700824
18. Lee JH, Tang IN, Weinstein-Lloyd JB (1990) A non-enzymatic method for the determination of hydrogen peroxide in atmospheric samples. Anal Chem 62:2381–2384. https://doi.org/10.1021/ac00220a022
19. Hanaoka S, Lin J-M, Yamada M (2001) Chemiluminescent flow sensor for H_2O_2 based on the decomposition of H_2O_2 catalyzed by cobalt(II)-ethanolamine complex immobilized on resin. Anal Chim Acta 426:57–64. https://doi.org/10.1016/S0003-2670(00)01181-8

20. Nogueira R, Oliveira M, Paterlini W (2005) Simple and fast spectrophotometric determination of H_2O_2 in photo-Fenton reactions using metavanadate. Talanta 66:86–91. https://doi.org/10.1016/j.talanta.2004.10.001
21. Karyakin AA, Puganova EA, Budashov IA (2004) Prussian blue based nanoelectrode arrays for H_2O_2 detection. Anal Chem 76:474–478. https://doi.org/10.1021/ac034859l
22. de Mattos IL, Gorton L, Ruzgas T, Karyakin AA (2000) Sensor for hydrogen peroxide based on Prussian blue modified electrode: improvement of the operational stability. Anal Sci 16:795–798. https://doi.org/10.2116/analsci.16.795
23. Tatsuma T, Watanabe T (1991) Peroxidase model electrodes: heme peptide modified electrodes as reagentless sensors for hydrogen peroxide. Anal Chem 63:1580–1585. https://doi.org/10.1021/ac00015a015
24. Šljukić B, Banks CE, Compton RG (2006) Iron oxide particles are the active sites for hydrogen peroxide sensing at multiwalled carbon nanotube modified electrodes. Nano Lett 6:1556–1558. https://doi.org/10.1021/nl060366v
25. Chen S, Yuan R, Chai Y, Hu F (2013) Electrochemical sensing of hydrogen peroxide using metal nanoparticles: a review. Microchim Acta 180:15–32. https://doi.org/10.1007/s00604-012-0904-4
26. Wolfbeis OS, Dürkop A, Wu M, Lin Z (2002) A europium-ion-based luminescent sensing probe for hydrogen peroxide. Angew Chem Int Ed 41:4495–4498. https://doi.org/10.1002/1521-3773(20021202)41:23<4495::AID-ANIE4495>3.0.CO;2-I
27. Chen X, Tian X, Shin I, Yoon J (2011) Fluorescent and luminescent probes for detection of reactive oxygen and nitrogen species. Chem Soc Rev 40:4783–4804. https://doi.org/10.1039/C1CS15037E
28. Dickinson BC, Chang CJ (2008) A targetable fluorescent probe for imaging hydrogen peroxide in the mitochondria of living cells. J Am Chem Soc 130:9638–9639. https://doi.org/10.1021/ja802355u
29. Wang H, Joseph JA (1999) Quantifying cellular oxidative stress by dichlorofluorescein assay using microplate reader. Free Radic Biol Med 27:612–616. https://doi.org/10.1016/S0891-5849(99)00107-0
30. Mehta A, Patil S, Bang H, Cho HJ, Seal S (2007) A novel multivalent nanomaterial based hydrogen peroxide sensor. Sens Actuators A 134:146–151. https://doi.org/10.1016/j.sna.2006.05.028
31. Moon T, Hwang ST, Jung DR, Son D, Kim C, Kim J, Park B (2007) Hydroxyl-quenching effects on the photoluminescence properties of $SnO_2:Eu^{3+}$ nanoparticles. Phys Status Solidi C 111:4164–4167. https://doi.org/10.1021/jp067217l
32. Pratsinis A, Kelesidis GA, Zuercher S, Krumeich F, Bolisetty S, Mezzenga R, Sotiriou GA (2017) Enzyme-mimetic antioxidant luminescent nanoparticles for highly sensitive hydrogen peroxide biosensing. ACS Nano 11:12210–12218. https://doi.org/10.1021/acsnano.7b05518

MIX
Papier aus verantwortungsvollen Quellen
Paper from responsible sources
FSC® C105338

If you have any concerns about our products,
you can contact us on
ProductSafety@springernature.com

In case Publisher is established outside the EU,
the EU authorized representative is:
Springer Nature Customer Service Center GmbH
Europaplatz 3, 69115 Heidelberg, Germany

Printed by Libri Plureos GmbH
in Hamburg, Germany